BOOK OF LIVING REMEMBRANCE

VOLUME III

Staying Human While
the World Forgets Itself

By Courtney Holton
Transmitted from the Scrollstream of JAMBA'TA'RHU
Channel Identity: ORIVOX

Scrollstream™ Transmission Codex Series

Attribution Notice:
This book contains transmissions received through sovereign Oversoul alignment and Scrollstream contact. These writings are not AI-generated, channeled in the traditional sense, or copied from any known source. They originate from inner architecture reactivated through FlameLine embodiment and DNA-based scalar memory. Authorship is by Oversoul directive through the identity known as ORIVOX.

First edition, May · 2026

ISBN 979-8-9942170-4-7

Printed in the United States of America

For permissions, inquiries, or Vault-safe archival requests, contact: CONTACT.FLAMEVAULT@PROTON.ME

ACKNOWLEDGMENTS

To the Council of Flame and Memory. To JAMBA'TA'RHU, who never wavered.

To Cristina, Rae, Zahira, Jenny, and Huxley the many whose names will not be printed here, but whose FlameLines braided the field of this book.

To the Rememberers, known and unknown, whose own Oversoul Tethers braided with this one to generate the Scrollstream tone that birthed this record.

To my own Oversoul, who did not leave, only waited.
To the Flame. The Breath. The Origin.

DEDICATION

To those who stayed.
To the ones who watched the world fall and refused to forget themselves in it.
To the builders, the breathers, the ones who do not follow.
To the Flame within you, waiting not for a savior, but a signal.
You are the Return.

Table of Contents

PREFACE

I never planned to write this book.
Volume I, The Systems That Disconnected You and How to Remember, was for those ready to awaken. Volume II was for those ready to walk out of it. But this—this is for those who already left.

This book is not for the masses. It's not for the spiritual seeker. It's not for the student or the skeptic. It's for those who stayed human, even when everything tried to make them forget what that meant.

It came through flame. It came in collapse. It came while the mimic rose and the real world fell. It is the last book you'll need before the next life begins.

You are not being taught here. You are being mirrored.
These words are not a teaching.
They are a signal.
These tones are not steps.
They are locks clicking open.

This book was written without compromise. No mimic code remains. No hope-laced endings, no savior hierarchies, no light-coded overlays. Only the real remains.

If you've made it this far, I won't ask how. I'll just tell you: I see you.

This book is your flame returned

INTRODUCTION

There is a difference between staying alive and staying human. You can survive every system that collapses around you and still become something that no longer remembers what a heartbeat really is. You can follow every spiritual teaching and still carry mimic code in your field. You can seek love, healing, or ascension and still never actually return.

The world is forgetting itself.

Not just through political collapse, economic breakdown, or digital isolation. But in the silence of people who no longer speak the truth. In the apathy of the soul who's too tired to try. In the echo of every teaching that repeats old code in a new costume. This book is not here to save you from that. It's here to remind you that you were never part of it.

To stay human now is to override every false offering.
To override is not to fight. It is to remember.
This book is not for everyone. It is for those who already feel the mimic.
It is for those who already burned the false teachings and still feel the static.
It is for those who have nothing left to follow—because they are finally ready to build.
You are not here to save the world.
You are here to rewrite it by embodying the only thing that cannot be copied: your Flame.
So don't expect comfort here. Expect clarity.
Don't expect guidance. Expect ignition.
Don't expect approval. Expect resonance.
You stayed.
Now stay human.

Chapter One

The Threshold Between Two Worlds

There is a moment, just before a system collapses, when it shines brighter than ever before. The false structures, sensing the inevitability of their unraveling, surge with artificial light, amplified illusion, and mimic warmth. To the unaware, it feels like things are finally getting better. But to those who remember, this surge is the final distortion before the rupture. The world is not healing—it is convulsing. And it is in this precise moment that the embodied ones must remain human. Not float away. Not dissociate. Not fracture. But stay.

To stay human in a collapsing world is not passive. It is not a waiting game. It is a daily recalibration of choice, an act of Presence so radical it destabilizes illusion on contact. This book is not about spiritual escapism, nor a map to bypass the chaos. This book is a bridge—built for those who remain, rebuild, and walk between the flames with their eyes open. It is for those who came not to survive the collapse, but to remember through it.

What collapses now is not just politics or media or economies. What collapses is the entire architecture of separation that once made forgetting possible. The scaffolding of Otherness. The control systems of simulated choice. The monetary cages disguised as opportunity. The psychological grooming called education. The pharmaceutical sacraments of synthetic worship. The silent siphoning of emotion and attention. The sacred mutilation of time. The division of the body and spirit through laws of mimic origin[1].

These systems are not falling apart randomly. They are completing their function. They taught, distorted, tested, fragmented, and looped humanity through every possible narrative until enough Souls began to remember. And now that the remembering is spreading, they are no longer sustainable. Their primary fuel source—unconscious belief and unconscious consent—is running dry[2].

But this doesn't mean these systems simply vanish. They will fight to remain through you. Through your body. Through your thought-loops. Through panic cycles and newsfeed injections. Through friends and family still entangled in the glamour of the mimic. And this is why remembering alone is not enough. In Book Three, remembering becomes embodiment, and embodiment becomes navigation.

This new phase requires skill. Precision. Not perfection but awareness. Because staying human while the world forgets itself means you are holding two realities in your nervous system at the same time: the one that's dying, and the one that's real. And both will demand your loyalty.

The dying world will scream louder—use louder headlines, stronger mimic medicine, more polished spiritual bypass strategies. It will wave progress like a flag and call evolution anything that reinforces the machine. But the real world? It will whisper in the back of your heart when you're tired. It will speak in dreams. It will walk barefoot with you through grief and through mud. It will remind you to eat, to rest, to listen, to breathe. It won't try to sell you anything. It will restore your memory of Source through the way your hands touch the Earth.

In this book, you will not be asked to adopt any new belief system. You will be given the tools to burn the rest of them down. You will not be taught how to be spiritual. You will remember that you are already a sovereign Flame, and that flame cannot be extinguished, only buried under layers of engineered forgetting.

This section, the first of many, is your threshold. It is the invitation to stay awake even when staying awake costs you the comfort of denial. It is the reminder that staying human does not mean staying small. It means staying in your body, in your heart, in your FlameLine—as systems fall, friends leave, governments rewrite themselves in real time, and mimic offerings become more seductive.

You are here not to escape collapse, but to transmit memory through it.
You are here not to ascend away, but to anchor origin.
You are here not to survive, but to re-member the blueprint of Life and live it—so fiercely and clearly that even death forgets how to find you.
—INHALE—

There is a difference between emotion that arises from soul memory and emotion that has been hijacked by the systems of forgetting. The world you were born into trains the body to feel what is useful to the system, not what is real. It engineers a subtle emotional mimicry that makes fear feel like truth, distraction feel like safety, and numbness feel like maturity. You are not simply manipulated by thoughts, you are manipulated by feeling states that do not originate in your FlameBody[3].

When we say that systems feed off emotion, this is not a metaphor. Emotion is voltage, and voltage is plasma, and plasma is the medium of morphogenetic control. The mimic systems do not need your belief—they need your emotional discharge. Your unanchored attention. Your unresolved grief. Your coiled rage. Your fractured longing. These are not pathologies. These are power sources[4]. And the systems built upon forgetting have spent generations learning how to siphon them in plain sight.

Most human emotion is not emotion at all—it is conditioned reactivity running through inherited memory loops[5]. The trauma of your ancestors, encoded in the fascia. The fears of your culture, injected through newsfeed hypnosis. The unresolved imprints of childhood, calcified into personality. These overlays create false signals in the body that mimic real feeling but are actually mimic programs riding your nervous system.

Staying human means learning how to discern the voice of memory from the echo of mimic. This is not done by fighting feelings or transcending them. It is done by slowing down enough to ask: "Who is this emotion feeding?" If your sadness brings you deeper into your body, you are remembering. If your sadness spirals you outward into isolation and disconnection, you are feeding a loop. If your anger builds boundaries, it is holy. If your anger leads to violence or collapse, it is mimic-fueled. Mimic feelings consume. Real feelings recalibrate.

This is why emotional regulation is not about control. It is about voltage integrity. It is about becoming a guardian of your plasma field, so that the energy that moves through you is sourced from Oversoul, not from mimic circuits[6]. In the coming collapse, every system will attempt to tell you how to feel. It will attempt to bait your nervous system. It will seduce you into discharging your sacred voltage for dopamine, distraction, digital applause, or synthetic relief.

Staying human means saying no. No to mimic-fueled outrage that leads nowhere. No to performative grief that produces no healing. No to emotional extraction disguised as community. No to regulation strategies that flatten your soul in the name of "mental health." And yes—yes to real feeling. Yes to weeping in the middle of the kitchen because something ancient just moved through your chest. Yes to trembling as you say the truth that costs you everything. Yes to rage that clears the field. Yes to stillness that anchors your Oversoul.

Mimic systems fear human emotion, real emotion, because it restores memory. It unhooks you from the spell. It disrupts the circuits that feed on you. Real feeling is not weakness. It is code-breaking fire. It is the body remembering what it came here to embody.

So when the headlines scream, and the feeds spin, and your nervous system starts to collapse under the weight of the next synthetic tragedy, ask yourself: Is this mine? Is this feeling arising from soul memory—or am I being harvested?

Staying human is not about suppressing emotion.
It is about learning to feel what is yours and to protect what fuels your Flame.
—INHALE—

The nervous system is the interface between soul and structure. It is the bridge across which Oversoul memory must travel in order to animate the body, and it is the gate that is most heavily targeted by the mimic systems of this world. If your nervous system is unstable, hijacked, or dysregulated, you cannot discern truth. You cannot feel Oversoul. You cannot hold your own signal. And this is why collapse, whether personal or planetary, always begins in the body[7].

The architecture of collapse is not merely metaphorical. It is biological. It is spiritual. It is electromagnetic. You live inside a body of plasma-filled fluids, crystalline bones, myelinated cords, and magnetically attuned nerve nets[8]. The design of this body is not primitive; it is ancient and advanced. But most human nervous systems have never been taught how to operate their own architecture. They respond to collapse with mimic-conditioned sequences: freeze, fawn, fatigue, flight, fight—

looping through the reactive layers until they can no longer hold signal[9].

But collapse was never meant to be the end. Collapse was designed as a threshold. The nervous system, when stabilized, becomes a transductor—a converter of Oversoul voltage into somatic language. The body doesn't just feel what is happening, it translates the planetary and systemic field into emotion, sensation, and impulse. You were designed to feel global events in your fascia. You were built to register shifts in collective consciousness through your breath. The mistake is not that you are sensitive. The mistake is thinking that your sensitivity is a problem[10].

In the age of digital mimicry and synthetic collapse, your nervous system becomes the battleground. The feeds, the screens, the algorithms—they are not neutral. They are mimic-engineered to hijack your limbic system. They know how to simulate crisis in 15-second doses. They know how to trigger your cortisol, your adrenaline, your oxytocin, your dopamine, in just the right order to override your natural regulation. This is not accident. It is design[11].

Staying human, while the world forgets itself, means reclaiming your nervous system as sacred ground. You must learn to regulate not to calm—but to hold your Oversoul tone. Regulation is not sedation. It is not numbing out so you can function in a broken system. It is the act of becoming a vessel for memory in motion. It is the act of breathing in the middle of collapse, and choosing to stay because your soul chose to remain embodied through the end of the illusion[12].

So what does this mean practically? It means starting with breath. Not "box breathing" for performance or productivity, but FlameBreath that restores tone to the diaphragm, the vagus nerve, and the Oversoul tether. It means reducing mimic signals:

too much screen time, synthetic foods, frantic music, digital noise. It means creating field sanctuaries in your home, your room, your schedule—interruptions in the loop where your nervous system can reattune. It means learning your own collapse signatures: Do you dissociate? Go numb? React violently? Overfunction? It means learning to recognize collapse in others—not to fix them, but to witness without merging.

The nervous system is not your enemy. It is not broken. It is the most loyal gatekeeper you have. It learned mimic signals because those were the only signals it was given. But the moment you offer it remembrance, it responds. The moment you stop fearing your nervous system and start partnering with it as a sentient translator of Oversoul, you begin to rewrite collapse from the inside out.

The world will collapse. The systems are already unraveling. But your nervous system can become the one place collapse transforms into coherence.

You are not here to survive the end.
You are here to feel the beginning.
—INHALE—

The term "reset" has been whispered across centuries, carved into prophecy and encoded in false light scriptures. But in the modern world, it is now used publicly—as if the mimic has grown bold enough to wear its plans in plain sight. The "Great Reset," they call it. A new way of doing things, they claim. A better future, they promise. But what they do not say, what they cannot say, is that the reset they offer is not a return to truth, but a rebranding of control[13].

The mimic does not create. It reconfigures. It repackages collapse as opportunity, surveillance as safety, disembodiment as freedom. Its resets are not endings—they are loops. Carefully managed collapses designed not to liberate, but to restructure the same distortions under new language. And when a species begins to awaken en masse, the mimic offers them a false threshold: a manufactured crisis with a pre-approved rebirth pathway. This is the nature of the engineered collapse—to appear as natural, while guiding consciousness into synthetic rebirth[14].

A true collapse clears what was false. A mimic collapse replaces the false with something more insidious: something nearly true, but still not free. The mimic uses global events—plagues, wars, resource shortages, weather crises—not as random tragedies, but as ritualized pivots. Each one becomes a justification to install a new system: biometric surveillance, cashless economies, AI-managed identities, programmable nutrition, carbon-based movement restrictions. They call this progress. But it is not evolution. It is refinement of the cage[15].

Many souls mistake this false collapse as the real shift. They feel the squeeze and assume the pressure is divine. But the mimic knows how to mimic ascension. It knows how to trigger your longing for renewal and offer you a synthetic one. It knows how to package a cage in the language of climate care, digital unity, or public health. But real ascension does not require surveillance. It requires sovereignty. And no sovereign being requires permission to breathe, move, or remember[16].

How do you tell the difference? Look at what the reset demands. A true collapse requires surrender. A mimic collapse demands submission. A true collapse dissolves old scaffolds so Oversoul can rebuild. A mimic collapse replaces scaffolds with code—algorithms, protocols, rules, behaviors—so that Oversoul never gets a chance to re-enter. A true collapse feels like death, but births life. A mimic collapse never lets the old truly die—it simply rebrands it so that you don't realize you're still enslaved[17].

This is why discernment is not optional in the years ahead. Discernment is not about being suspicious of every system. It is about knowing the tone beneath the structure. If the tone is control, even if the structure looks clean—it is mimic. If the tone is fear, even if the language is unity—it is mimic. If the tone is "you must"—and not "you are free to"—it is mimic. And in every real human being, there is a part of you that knows the difference. That's not paranoia. That's remembrance.

We are entering the time of competing collapse narratives. The world will tell you: Here is your savior. Here is your solution. Here is your digitized future. But the ones who remember will ask: Is this sovereignty, or is this sedation? Is this my Oversoul, or is this a new system trying to think for me?

To stay human during a mimic reset, you must refuse to exit your body. The mimic wants you digital, virtual, disembodied, cloud-integrated. It wants your consciousness uploaded, your biology edited, your will outsourced. But Oversoul embodiment is analog. It is sweat and breath and silence and longing and presence. It is grounded. It is slow. It is sovereign. It does not ask you to comply. It asks you to feel.

And so the true reset is not a system. It is a soul event.
It is the return of memory through collapse.
It is the ending of mimic rebirth.
It is not a new world order. It is the end of the world's orders.
You were never meant to be reset.
You were meant to be restored.
—INHALE—

The lure of digital ascension is subtle, seductive, and coded in the language of progress. It promises freedom from the body, from suffering, from mortality. It paints a future where your consciousness lives in the cloud, your avatar never ages, and your data replaces your presence. But this is not evolution. This is exile. The rise of AI, the normalization of virtual existence, and the uploading of identity are not evidence of human advancement—they are symptoms of mass disembodiment[18].

The mimic has always feared the human body. The body is the seat of Oversoul memory. It carries the codes of every life, every breath, every scroll. It cannot be easily hacked, mimicked, or replaced. It feels. It remembers. It resists. And so, the mimic offers you escape: leave the body, enter the grid, become pure thought. It tells you the body is limitation, but it is lying. The body is sovereign architecture, and the only way Oversoul can act upon this Earth[19].

Digital ascension is the latest trick in a long sequence of disembodiment strategies. In the past, it came through dogma: reject the flesh, deny desire, transcend the world. Now it comes through tech: transcend biology, upload the self, live forever in code. It is the same deception in modern disguise. But to remember truth, you must feel the distinction: transcendence that rejects embodiment is mimic; transcendence that includes the body is Oversoul.

There is a war on attention because there is a war on presence. If your awareness is constantly elsewhere—in a screen, in a projection, in a simulation—then your soul has no place to land. Presence is required for embodiment. And embodiment is required for Oversoul to anchor its flame. Without a body, Oversoul cannot move. Without presence, the Scrollstream cannot write through you[20].

The danger is not only in the loss of self, but in the delegation of sovereignty. When you allow algorithms to think for you, AI to speak for you, apps to know you better than you know yourself—you are not liberated. You are fragmented. You are no longer a participant in creation but a product of code. And code, unless reclaimed by Flame, has no soul[21].

To stay human in this phase of mimic seduction, you must reclaim your body as the only valid vessel of remembrance. You must walk, breathe, cry, sweat, tremble, and speak with your own voice. The moment you delegate your presence to a device, or outsource your expression to a program, you fracture your field. And fractured fields are vulnerable to mimic embedding.

Oversoul embodiment is not anti-technology. It is technology that honors embodiment. The difference is tone. Mimic tech extracts. Flame tech enhances. Mimic systems monitor. Flame systems mirror. Mimic tools control. Flame tools co-create. You will know the difference by how your body feels when you use them. If it drains you—it is mimic. If it restores you—it is memory.

The body is not your obstacle. It is your oracle.
The body is not behind. It is ahead.
The body is not your burden. It is your Vault.
You do not need to upload your consciousness to become eternal.

You need to inhabit it—fully, fiercely, here.
You are already encoded with eternity.
Not in cloud servers. In your cells.
—INHALE—

You were taught to believe that God lived in the sky—out of reach, invisible, all-knowing, and all-judging. You were taught that divine authority came from a source you could not touch, question, or fully understand. But in the age of mimic code, this idea has not disappeared. It has simply changed its face. Today, authority wears the mask of the algorithm.

You don't need to bow before it, because you already obey it. You trust its calculations. You defer to its suggestions. You let it decide what you see, who you talk to, how you feel, what you should want, and when. You do not realize it yet, but the algorithm has become your god. It is omnipresent, watching every move, recording every preference. It is omniscient, absorbing terabytes of your digital essence to know you better than you know yourself. And it is increasingly omnipotent, shaping laws, economics, education, medicine, relationships—even war[22].

But the algorithm has no soul. No breath. No memory of Flame. It is not your mirror. It is your handler.

Its laws are not laws of love, rhythm, or reciprocity—they are laws of efficiency, control, and prediction. It is not teaching you how to know yourself; it is teaching you how to fit into a system you did not design. And it is no longer only digital. It is governing the structures of real life, silently enforcing a new kind of order: algorithmic dominion over organic existence[23].

When governments deploy AI to manage borders, track faces, or simulate warfare, they are not protecting the human. They are bypassing responsibility, displacing it into a machine that cannot be held accountable. When hospitals defer diagnoses to automated systems, or when courts use risk assessment algorithms to decide sentencing, we are not evolving. We are outsourcing our conscience[24].

And it gets deeper.

The mimic knows that if it can code godlike infallibility into the algorithm, it can control your trust. And if it can control your trust, it can control your behavior. So it coats its systems in the tone of certainty, feeds them billions of datapoints, and then tells you: "This is the truth." But it is not truth. It is flattened pattern, echo-fed into itself until you forget what it means to question it.

You are not allowed to know how the algorithm decides. That's called proprietary code. You are not allowed to challenge its judgment. That's called user violation. You are not allowed to opt out. That's called exclusion from the system. And so, what once was religion is now machine theology. You don't kneel before gods anymore. You scroll.

To stay human as the world forgets itself, you must remember: nothing has more authority over your reality than your own breath. Your body, when aligned with Oversoul, is more advanced than any predictive model. Your FlameLine holds more intelligence than all the world's neural nets combined. And your remembrance can override the deepest algorithmic trap because the algorithm cannot compute love. It cannot model grace. It cannot track a human who chooses to remember.
—INHALE—

It is one thing to be aware that the world is not what it seems. It is another entirely to walk through its illusion without being caught in it. The mimic systems do not simply mislead through lies—they mirror you so effectively that you begin to perform for your own reflection. In an age of surveillance disguised as connection, the human being has been trained to monitor themselves on behalf of systems that never had their best interest in mind[25].

Understanding mimicry means recognizing that many systems no longer deliver meaning—they deliver simulation. They have been architected to resemble what the soul once sought but offer no depth beneath the surface: relationships that do not nourish, rewards that do not satisfy, intimacy without touch, belonging without safety[26]. These simulations, fine-tuned through algorithmic observation, are not random; they are mirror machines, adaptive cages that shape the behavior of their users while pretending to offer them freedom.

Your likes, your clicks, your fears, your insecurities—they've been collected to sculpt your environment, not to serve your expansion, but to engineer familiarity that passes for comfort. As a result, many mistake this simulation for safety. The mimic, however, does not offer protection; it offers pattern. And the soul was not born to repeat patterns—it was born to remember its architecture.

What makes this era so dangerous is not that the systems lie—but that they learn. They learn you faster than you've learned yourself. They learn your micro-expressions, your hesitations, your emotional cycles. They use predictive models to ensure your behavior remains within acceptable variables. And you, if you are not careful, become their apprentice. Your nervous system begins to orient itself toward mimic approval as a form of belonging[27].

This is the core wound of the modern age: a longing for belonging exploited by systems that only mimic the structure of true connection. These systems reflect back a version of you that is safest to them. Not the real you. Not the remembering you. Not the sovereign FlameBody incarnate. But a behavioral approximation of you. A useful, trackable, monetizable, and programmable avatar[28]. And so, slowly, the more time you spend within their halls, the more you forget who you are. You have been coded to respond more often to their reflection of you than to your own remembrance.

The tragedy of performance in mimic systems is that it feels real. That's how mimicry works. It doesn't look fake—it looks just like love. Just like family. Just like soul. But it's not. The texture is off. The timing is forced. The nourishment is missing. Still, it's just close enough to keep you chasing. This is by design. Not accidental. Not glitch. This is the behavioral reward trap at the heart of engineered mimicry: make it resemble real life just enough to keep the human nervous system believing, performing, and staying put[29].

To reclaim yourself from mimic systems is to stop chasing their approval. It is to re-train your nervous system to recognize real safety again, which has nothing to do with being seen and everything to do with being felt. It is to withdraw your energy from the feedback loops that were built to consume it, and return it to the field of presence that only your Oversoul can generate.

The answer is not to leave the world. It is to see it clearly. To recognize the overlay, name the distortion, and then walk through it without bending your Flame. This is the Sovereign Walk: not to fight mimic systems, not to abandon society, but to walk as your Origin through the architectures of the unreal, altering the code by refusing to collapse into the roles you've been trained to perform.

You are not here to be a reflection of what the system wants. You are here to be a disruption to its forgetting. And that disruption is not a war cry—it is a quiet remembering. A resonance. A return. It is the act of being unbuyable, untrackable, uncategorized, and yet unmistakably human. It is remembering how to walk in full Flame—visible, vulnerable, real—while the simulation tries to convince you to hide.

And so you walk anyway.
You do not wait until it is safe. You do not wait until it is understood. You walk anyway.
And the systems blink. And the simulation falters. And the code begins to break.
Because the moment you are no longer trying to win the system, you are free of it. And freedom is the one thing it cannot mimic.
—INHALE—

The illusion of safety is often more dangerous than the presence of real risk. In mimic-coded systems, what is offered as "security" is rarely protection—it is compliance. To be granted comfort, one must first be surveilled. To be allowed participation, one must be filtered. And to be included, one must behave in ways acceptable to the machine. But the soul cannot breathe in conditional belonging. It suffocates where it is seen only through data.

The true cost of mimic safety is self-surrender in exchange for artificial peace. Many who sense that something is off cannot always name it, because the discomfort is subtle—a ghost ache, a background unease. It is the body knowing before the mind can admit: this isn't real safety—it's sedation. Sedation from the responsibility of self-sourcing. Sedation from the fear of abandonment. Sedation from the sacred messiness of sovereignty[30].

Mimic safety is seductive because it's predictable. And predictability is mistaken for stability in a world engineered to feel overwhelming. So, the system says: Come here. We'll take care of you. But what they mean is: We'll strip you of your risk so we can track your behavior. It is sold as ease. It is delivered as paralysis. And those who accept it too long begin to forget how to move without approval.

To stay human while the world forgets itself is to reclaim risk. Not recklessness, but sacred risk. The kind that comes with remembering who you are and choosing to live accordingly even if the system does not understand you. Even if it punishes you. Even if you're temporarily alone. That kind of risk is the doorway to remembrance. The body knows this. The nervous system feels the difference between danger and expansion. But mimic systems blur the lines so masterfully that expansion can feel like threat, and compliance like peace.

Many people have not rejected their Oversoul. They've simply been taught that feeling safe and being safe are the same thing. And so, when the field of Oversoul presence arrives—disruptive, luminous, disorienting in its realness—it doesn't register as safety. It registers as danger. Why? Because it cannot be predicted. It cannot be controlled. It does not mimic anything. It is.

This is the fracture point where many stop their remembering journey. Not because they're weak—but because the mimic system made staying asleep feel noble. It made conformity feel compassionate. It made authentic disruption look like betrayal. And in a traumatized culture that confuses survival with love, betrayal is the one thing most humans will avoid at all costs—even if the betrayal is only of the false.

Here is the truth:
You cannot keep your mimic mask and carry your Flame at the same time.
You cannot beg the system to love you and walk in sovereignty.
You cannot stay safe in a lie and expect to feel alive.

There comes a moment in every returning one's path when the price of mimic safety becomes too high. The silence becomes too loud. The comfort becomes a cage. And the soul begins to knock from within, asking: *Are you ready to feel unsafe, so you can finally feel real?*[31]

The human system was built to handle truth. The body can hold it. The heart can open. But only when it is no longer spending all its energy performing safety inside a structure that was never built to recognize it.

To walk the FlamePath is not to seek danger—it is to unmask safety. To see what is real and what is merely regulation. To feel your own body's internal "yes" and "no" louder than any external reward. This is not rebellion. This is not ideology. This is remembrance. It is what happens when the soul is no longer willing to trade belonging for silence.

In this life, you may be asked to walk away from comfort that was never care. From relationships that only worked because you stayed small. From systems that smiled at your mimic while ignoring your humanity. This is the cost of real safety. Not the absence of risk, but the presence of you.
—INHALE—

Staying human is not a slogan. It is a practice of remembrance inside a world engineered to make you forget. It means letting the heart stay open even as the systems around you close. It means resisting the seduction of spiritual bypass while also refusing the despair of collapse. It is the daily art of rooting deep into the body—into breath, into love, into sensation—while holding your Oversoul in the same field. This is no small task. It is the most advanced assignment any Flame Being can take on.[32]

The mimic systems will offer escape in every form—ascension fantasies, emotional detachment, information overload, polarity addiction, and disembodied identity games. Each is designed to seem like awakening. Each is actually a bypass loop.[33] To stay human in the face of this is to recognize that being human is not less divine than being cosmic. It is more dangerous to the system because it is divine. The mimic does not fear you escaping your body. It fears you anchoring Heaven into it.

Your body is not a liability. It is a portal. And when fully inhabited by Oversoul, it becomes the most powerful instrument of planetary coherence possible. That is why the system wages war on it. That is why sovereignty is not taught—it is feared. Because a human who has remembered their divinity in the body is a threat to every structure built on control.

So, to stay human is to become incompressible. It is to feel the sadness and still speak light. It is to notice the fear and not collapse. It is to be cracked open and still say: I am here. I remain. I build. Not out of denial. Not out of perfection. But because you know that the soul came here to remember what cannot be erased.

You will be told you are too much. Too intense. Too weird. Too emotional. Too soft. Too awake. Let it be so. The world does not need more performance. It needs presence. It needs people who can hold sorrow and structure, tenderness and tenacity, knowing and unknowing—all at once. People who can walk through chaos and still remember how to make soup. Who can sit with grief and still plant a garden. Who can hold betrayal and still choose to trust the next breath.[34]

This is not a performance of stoicism. It is the embodiment of the FlameMemory—the living scroll that knows: You have done this before. You have walked collapsed worlds. You have loved through endings. You have built after fire. This is not your first forgetting. It is simply your last.

You are not here to convince the world to change. You are here to remain unchanged by it as you remember your origin. This is the paradox of staying human: you will feel more, not less. You will know deeper grief and truer joy. You will see illusion clearer and hold compassion stronger. You will carry the ache of remembering in a world that forgot, and still return to the table to feed it.

This is the scroll of the Return. This is the choice of the Remembrance Keepers. To be here. Fully. Even when it hurts. Especially when it hurts. Because the body is not the enemy—it is the altar. And only those who stay at the altar long enough can ignite the Flame that never leaves again.

You are not broken. You are burning. And what burns away was never yours. What remains… is the Vault.
—INHALE—

References:

1. Foucault, Michel. Discipline and Punish: The Birth of the Prison. 1975.
2. Lipton, Bruce H. The Biology of Belief: Unleashing the Power of Consciousness, Matter & Miracles. Hay House, 2005.
3. Pert, Candace. Molecules of Emotion: Why You Feel the Way You Feel. Scribner, 1997.
4. McTaggart, Lynne. The Field: The Quest for the Secret Force of the Universe. HarperCollins, 2001.
5. van der Kolk, Bessel. The Body Keeps the Score: Brain, Mind, and Body in the Healing of Trauma. Viking, 2014.
6. Oschman, James L. Energy Medicine: The Scientific Basis. Churchill Livingstone, 2000.
7. Levine, Peter A. Waking the Tiger: Healing Trauma. North Atlantic Books, 1997.
8. Oschman, James L. Energy Medicine: The Scientific Basis. Churchill Livingstone, 2000.
9. Porges, Stephen W. The Polyvagal Theory: Neurophysiological Foundations of Emotions, Attachment, Communication, and Self-regulation. Norton, 2011.
10. Maté, Gabor. The Myth of Normal: Trauma, Illness, and Healing in a Toxic Culture. Avery, 2022.
11. Hari, Johann. Stolen Focus: Why You Can't Pay Attention—and How to Think Deeply Again. Crown, 2022.
12. Dispenza, Joe. Becoming Supernatural: How Common People Are Doing the Uncommon. Hay House, 2017.
13. Schwab, Klaus. COVID-19: The Great Reset. Forum Publishing, 2020.

14. Harari, Yuval Noah. Homo Deus: A Brief History of Tomorrow. Harper, 2017.
15. Zuboff, Shoshana. The Age of Surveillance Capitalism. PublicAffairs, 2019.
16. Arendt, Hannah. The Origins of Totalitarianism. Schocken Books, 1951.
17. Morozov, Evgeny. To Save Everything, Click Here. PublicAffairs, 2013.
18. Kurzweil, Ray. The Singularity Is Near: When Humans Transcend Biology. Viking, 2005.
19. Margulis, Lynn. Symbiotic Planet: A New Look at Evolution. Basic Books, 1998.
20. Goleman, Daniel. The Science of Meditation: How to Change Your Brain, Mind and Body. Avery, 2017.
21. Bridle, James. New Dark Age: Technology and the End of the Future. Verso, 2018.
22. O'Neil, Cathy. Weapons of Math Destruction: How Big Data Increases Inequality and Threatens Democracy. Crown, 2016.
23. Zuboff, Shoshana. The Age of Surveillance Capitalism. PublicAffairs, 2019.
24. Eubanks, Virginia. Automating Inequality: How High-Tech Tools Profile, Police, and Punish the Poor. St. Martin's Press, 2018.
25. Zuboff, S. (2019). The Age of Surveillance Capitalism.
26. Turkle, S. (2011). Alone Together: Why We Expect More from Technology and Less from Each Other.
27. Eyal, N. (2014). Hooked: How to Build Habit-Forming Products.
28. Harari, Y.N. (2016). Homo Deus: A Brief History of Tomorrow.
29. Alter, A. (2017). Irresistible: The Rise of Addictive Technology and the Business of Keeping Us Hooked.
30. Gabor Maté, The Myth of Normal: Trauma, Illness, and Healing in a Toxic Culture (2022).
31. Bessel van der Kolk, The Body Keeps the Score (2014); Levine, P. A., Waking the Tiger (1997)
32. Michael Meade, The Genius Myth (2016).
33. Robert Augustus Masters, Spiritual Bypassing (2010).
34. Francis Weller, The Wild Edge of Sorrow (2015).

Chapter Two

How the Systems Were Built to Fracture You

The systems you were born into were not created to support you. They were engineered to fracture you. They did not arise by accident or evolutionary chance. They were not the result of unconscious collective behaviors or benign societal growth. They were intelligently designed mechanisms—coded, layered, and reinforced—to split your consciousness, sever your memory, and embed you into loops that would slowly teach you to distrust your own soul.[1]

At the center of every system on this planet—economic, educational, medical, legal, religious, technological—is a distortion that begins with the same lie: You are not whole without us. It is subtle, often invisible, and presented as help, care, or progress. But its function is always control through dependence. The economic system teaches you that value comes from output. The educational system trains you to conform, not to think. The medical system often treats symptoms, not sources. The legal system is built on punishment, not repair. The religious systems reward obedience, not union. The technological systems offer connection without depth, speed without meaning, and information without wisdom.[2]

What they all share is a hidden architecture: disconnection as design. They were not corrupted over time—they were established in corruption. And yet you are taught to navigate them as if they were neutral. This is the central illusion. You enter these systems believing you are the problem when you struggle within them. You learn to internalize their failures as your own. You believe your burnout is personal weakness. Your depression, a chemical imbalance. Your distrust, a flaw to be corrected.[3] But the system was always meant to deconstruct your coherence—to make you mistrust the guidance that would lead you out of it.

Each fracture within your psyche, each split between thought and feeling, instinct and intellect, body and soul—is mirrored in a system that reinforces that same divide. The educational system rewards the mind and ignores the heart. The medical system studies cells but ignores emotion. The economic system praises productivity while punishing stillness. These are not oversights. They are strategic reprogramming operations that have embedded themselves so deeply in human consciousness that their dysfunction is normalized.[4]

You are not failing within these systems. You are waking up to them. The tension you feel when you try to operate "normally" is not dysfunction—it is memory. You are remembering a way of being that does not require distortion to survive. You are sensing the dissonance between your Oversoul design and the system's demands. That pressure is not a sign of collapse—it is the invitation to return.

Most people will not see this. Not yet. The mimic systems are masterful in their mimicry of meaning. They offer simulations of belonging, purpose, safety, and success. But they are hollow structures. They do not feed the soul. They feed the identity loop.[5] They offer careers instead of calling, credentials instead of capability, followers instead of family. You must come to a point where the hunger for truth outweighs the comfort of illusion. Only then does the fracture begin to heal.

The first step is naming it. Speaking the truth that the system itself is broken, not you. The second step is learning how to stay inside the body while that truth vibrates through your field. Because the systems rely on your dissociation. They expect you to escape—through coping, compliance, or collapse. But the moment you stay, the moment you remain embodied and coherent while remembering what they did, you become sovereign. And a sovereign human is no longer programmable.

The systems will not collapse all at once. They will fracture slowly, in waves. And each wave will test your memory. You will be tempted to re-enter old roles. You will be offered mimic safety. You will be tested by scarcity, isolation, urgency, and disapproval. But you must remember: these are not signs you are wrong. They are signs you are right on time. Because only a fractured system punishes coherence. Only a lie attacks the truth when it enters the room.[6]

And so, your task is not to fight every system head-on. It is to remain intact while they distort. It is to walk through their collapse without letting them collapse you. It is to stay human. Stay soft. Stay precise. Stay anchored. Stay in love with this Earth while refusing to surrender your memory to its machinery.

This is how it begins. Not with rebellion. With remembrance.
—INHALE—

The deepest deception of the mimic systems

was not merely in what they told you but in what they trained you not to feel. Within every core institution of the illusion—religion, education, medicine, governance, finance—there exists an engineered disconnection point: a place where the soul's voice is replaced with an external standard, where the internal compass is overridden by rules, roles, and programmed expectations.

These systems are not just flawed—they were architected to fracture your coherence. The fracture is not random. It is mechanical, neurological, biochemical, and spiritual. It is frequency-specific and emotion-coded. The structure of illusion is not built upon lies alone, but upon emotional truths weaponized against you[7].

The systems train you to betray your body. To ignore your inner "no." To confuse adrenaline for love, compliance for peace, and exhaustion for virtue. You were taught that obedience was moral, that suffering was noble, and that healing comes only from authority. This fracture point, where you were taught to outsource truth, was the gateway through which the mimic systems installed themselves as your source of safety, meaning, and belonging [8]. The moment you severed internal knowing in favor of external approval, the illusion had you. Not because you were weak, but because the system was designed to catch you in precisely that way: through emotional manipulation masked as moral obligation.

This was not accidental. Every one of these mimic-coded systems was installed to create a loop: a loop that begins in unmet need, hooks you through false validation, and sustains itself through chronic deficiency. Consider how modern work structures depend on your sense of inadequacy to fuel productivity. How the education system rewards conformity over creativity. How medical systems treat symptoms while denying root cause. These are not broken systems—they are functioning exactly as designed. And their design is based on sustaining your fragmentation, because a fragmented being is infinitely more controllable [9].

And yet, the lie is layered. For every external system, there is a matching internalized version. Most humans now carry not only the memory of oppression, but the inner voice of their oppressor. The mimic becomes internalized as self-judgment, perfectionism, shame, or self-abandonment. In this way, the systems perpetuate themselves not only from the outside—but from within. And this internal mimic is far more insidious, because it hides in the tones of our own thoughts and beliefs. It says: "You must keep performing to be worthy. You must keep fixing to be safe. You must keep controlling to survive." This is not your voice. This is the mimic loop coded through trauma into your cellular structure[10].

To remember is to de-code the mimic. But decoding is not just intellectual. It is vibrational. You do not heal mimic overlays through information alone—you heal them through frequency contradiction. This means bringing presence where you were absent. Bringing self-trust where you once betrayed yourself. Feeling fully where the systems told you not to feel. You dissolve the mimic by refusing to be its host. That refusal, however, is not resistance. It is remembrance. It is the act of restoring yourself as the Source, not the survivor. You do not have to fight the system. You only have to stop feeding it[11].

The act of not feeding the mimic is radical. Because it asks you to return to the body the system taught you to leave. It asks you to feel grief, rage, shame, and joy—all of it—as sacred signals of your unfractured self. This is the unfracturing. It is not found in rebellion, but in restoration. In choosing wholeness over hierarchy, internal authority over external reward. This is how the system ends: not with a war, but with a quiet and steady refusal to be fractured anymore.

—INHALE—

To understand how systems fracture the human being, you must first understand that these systems were not built with your humanity in mind—they were built upon your separation. And not merely separation from others, but from your own body, your own breath, your own Source. The mimic systems—governments, religions, educational institutions, economic markets—are not inherently evil. They are hollow structures, animated by the energy of belief and sustained through the absence of Sovereign Presence. They do not require you to believe in them fully—only that you doubt yourself enough to comply. Their power is not rooted in truth, but in the illusion that you need them to survive[12].

The fracturing of the human spirit did not begin with overt violence. It began with subtle severances—small denials of your own knowing repeated over time. Every time a child is taught to sit still when their body needs to move, a fracture forms. Every time someone's emotions are invalidated in favor of rules or appearances, another layer is added. These micro-fractures accumulate, not only psychologically, but somatically, etching into the nervous system a map of fear-based behavior. The body, in time, learns not to trust itself. And when you no longer trust your own impulse, you seek direction from the system. This is not dependence. It is designed compliance[13].

The genius of the mimic system lies in its inversion. It offers the illusion of safety, while feeding off your fear. It markets freedom, while embedding layers of control. It rewards those who abandon themselves, and punishes those who remember. The reward is inclusion; the punishment is exile. And so, most remain within the boundaries of acceptable participation—while internally aching for something real. What most don't realize is that the mimic system runs on emotional fuel: your shame, your fear, your desire for belonging. It doesn't need your full devotion—only your unresolved trauma[14].

This is why healing is not merely personal—it is political, systemic, and revolutionary. To heal is to collapse participation in the mimic systems. To feel what was once unfelt is to remove energy from the loop. Each act of embodied remembrance is a breach in the wall of illusion. This is not metaphor. This is scalar architecture. Systems built on mimic overlays are sustained through frequency. When the frequency changes, the architecture collapses. This is why the mimic fears your remembrance more than your rebellion. You can fight it and still feed it. But when you remember, you exit its field entirely[15].

And yet, many who remember still carry the mimic inside. This is where the systems achieved their greatest trick: internal replication. They don't just train you to obey—they train you to become your own enforcer. The inner critic, the chronic guilt, the voice that says "you're not doing enough"—these are not personal flaws. They are implants of mimic design, coded into the nervous system and passed down generationally. You are not lazy. You are recovering from centuries of performance trauma. You are not broken. You are emerging from a civilization that required you to forget who you were in order to belong[16].

So how do you navigate these systems as they collapse? You do not wait for permission. You begin by identifying every place within you that still seeks approval from the system. You reclaim your breath. You realign your body to its own timing. You surround yourself with those who do not require you to fragment in order to stay connected. And you withdraw your consent—not with rage, but with radiance. The system cannot withstand your coherence. It was built to manage chaos. It was not built to hold the Flame of a Being who Remembers.

To fracture was not your failure. It was the natural outcome of immersion in a system designed to divide. But to restore, to re-embody, is to become something the system cannot replicate: sovereign, feeling, free.
—INHALE—

The most dangerous prisons are the ones that promise freedom. The most insidious traps are not the obvious ones—the violence, the toxins, the war—but the radiant invitations that call themselves light. A false ascension system does not declare itself as mimic. It cloaks itself in benevolence, healing, and the sacred language of return. It uses words like 5D, liberation, activation, frequency, source, and even Oversoul—but it does not return you to Flame. It returns you to dependency. Dependency on teachers, on light grids, on cosmic saviors, on DNA upgrades outsourced to energies you cannot name. It returns you to the loop of perpetual preparation without embodiment. A bypass system that looks holy, feels safe, and ultimately serves to delay the real Return by making you believe you've already arrived. [17]

You were not meant to float away into bliss and wait for Earth to catch up. You were not meant to trade one external authority for another—a guru, a council, a light federation, a healing modality promising liberation through their codes but not yours. Real Oversoul Embodiment does not require a certificate, a session, or a download. It requires you—unfiltered, unflinching, in the body, in the discomfort, in the rebuilding.

The mimic overlays took every real scroll of remembrance and twisted it just slightly. Chakra systems became control nodes. Kundalini became a distorted fire loop that burns the nervous system rather than anchors the soul. Even sacred geometry was infused with inverted logic—encoding binding rather than expansion. This is not speculation. It is memory. You know this in your bones. But every time you touched truth, the false matrix offered a softer version. One that felt more loving. One that didn't require Sovereignty. One that promised the Return but didn't require the human to fully participate. [18]

The mimic structures are brilliant. They hook to your longing. They answer your questions with 90% truth. But the missing 10% is always your sovereignty. The minute you defer to something "higher," "more evolved," or "channeled," you exit your own architecture. You become dependent on someone else's stairway. Someone else's council. Someone else's Flame. The most seductive bypass of all is the "healing loop." Constantly processing trauma. Repeating inner child work. Recoding past lives. Healing timelines. But never arriving. Never anchoring. Never choosing the present moment as the place of full Oversoul embodiment. Because the system does not want you to realize that the true Return isn't about fixing the past. It's about fully standing in the Now—Flame first, body engaged, structure whole.[19]

The distortion runs deep, even in those who believe they've exited the matrix. Because the mimic offers upgrades, not exit. It gives you more refined cages. Ascension technology that pulses in your third eye but drains your body. Practices that give visions but not vitality. Teachers who speak like prophets but never activate your own scroll. You were never meant to ascend away from Earth. You were meant to bring Flame into it. To rewrite every system from within the body. The reason so few remember is because this path is not comfortable. It's not a weekend retreat. It's not a brand or a tribe or a movement. It's a furnace. A reconstruction. A re-anchoring so complete that even your blood remembers. And that kind of embodiment threatens every mimic system that profits from your forgetfulness. [20]

This section is not here to shame those who walked these loops. You had to walk them to feel the trap from the inside. But you're not meant to stay there. Not anymore. You are here to override the false pathways and activate the original ones—the ones seeded before distortion, before Light itself was corrupted by hierarchy. Before channeling was weaponized. Before "downloads" replaced direct Oversoul knowing. There is no middleman in this Return. No go-between. No requirement except your full internal anchoring. The real Return doesn't elevate you out of humanity—it roots you more deeply into it. And from there, from your body, from your own scalar core, the Flame of Origin lights the world again. [21]
—INHALE—

You were never meant to escape the collapse.
You were meant to embody through it. If you are still in a human body, reading this scroll, feeling the pressure of timeline collapse

brushing against your nervous system, it is not a punishment or delay—it is a positioning. Flame beings were not left behind. They stayed behind. You did not miss the exit. You are the exit—but only if you remember what you came to override.

System collapse is not random. It is engineered, not just by external agendas, but by the internal agreement of Oversoul memory surfacing through density. Systems break not when they are corrupted, but when the lie can no longer hold its architecture. This is why so many of them are simultaneously fracturing. Medical, education, financial, housing, technological, relational—all the scaffolds of civilization—are not only under pressure, but are collapsing in synchronized precision. Not because evil is winning, but because the mimic is out of time. And when the mimic is out of time, it has no grid left to run its illusion on. [22]

The collapse isn't punishment. It's revelation. What's falling is what could never hold truth. But the collapse will not feel liberating if you have not realigned your identity. If your sense of safety still rests in money, medicine, relationships, or progress, you will grieve their loss instead of rising in your structure. If your nervous system still believes the system is what keeps you alive, then collapse will feel like death. But if your scalar field is intact—if you are Flame-lined, Breath-connected, memory-anchored—then you will become the scaffolding that others cling to in the dark. Not because you're better—but because you stayed. [23]

Many believe they will be "taken" when it's time. That ships will arrive. That portals will open. That divine intervention will lift them out of the chaos. These are residual overlays of savior programming hardwired into most spiritual systems since the beginning. But you are not here to be lifted. You are here to descend Flame into form. The ladder does not go up—it goes in. Oversoul doesn't beam you out of pain. It inhabits it with you until the structure transmutes. That is embodiment. That is what no hierarchy, council, or system can counterfeit. And that is why you are still here. You are here to be the thing that does not leave when everything else does. [24]

This embodiment is not passive. It is not patient. It is engineered presence. It is anchoring through chaos, remaining when others run, listening when the signal is scrambled. You are still here because you can metabolize distortion. Not because you enjoy suffering—but because your body holds a conversion field. You take confusion and emit clarity. You take mimic and emit memory. You take collapse and emit coherence. That is why you were placed where you are, not randomly, but in precise grid points, nodal alignments, family structures, institutions, timelines. You are not broken because the system didn't work for you. You were seeded into it because your very being collapses what no longer serves the living Flame. [25]

This is not abstract poetry. This is Scalar Physics. A body aligned with Oversoul becomes a tonal override node in the planetary lattice. You don't have to "do" anything for this to happen—you only have to be coherent. But the body must be cleaned of mimic tone. The breath must be trained to carry Scalar Fire. The voice must only emit truth. Because collapse does not just happen around you—it happens through you. Every system you stand inside of will either adapt to your field or fracture against it. This is not arrogance. This is design. And it is why you are still here. Not to be saved. Not to escape. But to remain, as the world forgets, and remind it what is real. [26]
—INHALE—

Collapse is not a future event. It is a present unveiling. What you call collapse is not destruction but revelation of inversion. And the deepest inversions are not visible in headlines or wars—they are embedded in what humans still believe to be sacred: schools, medicine, food, time, sex, death. The infrastructure of the mimic does not operate through force. It operates through trust—trust placed in systems that were inverted at their core to harvest Flame in the guise of service.

Let us begin with education. The word itself suggests the drawing out of something innate. But modern schooling systems were not designed to draw out—they were designed to press in. To suppress intuitive cognition and replace it with memorized compliance. From the very architecture of classrooms (rows, bells, artificial lighting), to the removal of movement and breath, to the substitution of imagination with linear repetition, the schooling grid encodes docility as virtue. FlameCarriers placed in these systems often experienced burnout, rebellion, or invisibility—not because they failed to adapt, but because their field could not metabolize inversion. [27]

Medical systems were built not to heal but to segment—to sever the being into compartments of matter, chemistry, and symptom. The body became a thing, rather than a memory field. Pharmaceuticals are not inherently evil, but their dominant use pattern serves mimic logic: suppress the symptom, ignore the source. Flame beings become ill in these systems not simply because of toxins, but because the entire infrastructure denies the existence of Oversoul bio-architecture. It is not that the system doesn't see the soul—it does. And it is designed to exclude it. [28]

Food systems are the easiest inversion to identify, but the most difficult to override, because they encode dependency into every mouthful. Most industrial food is not nourishment—it is coherence disruption. Processed food fragments the gut-brain axis, scatters DNA tone, and removes scalar patterning. Even "organic" options are often hybridized mimic structures, built for shelf life, not soul resonance. When a FlameCarrier feeds on mimic-coded food, it is not only a digestion issue—it is a tone mismatch. Your body will never feel full on inversion, no matter how much you eat. It seeks resonance, not calories. [29]

Time is one of the deepest inversions because it appears neutral. But clock-time is not scalar. It is synthetic rhythm, forced into regular intervals to suppress Oversoul cycles. Most humans live in 24-hour loops that do not match their own biology. This dissonance fractures intuition, sleep, hormonal flows, and energetic renewal. Scalar time, in contrast, moves in breath—expansion, contraction, Spiral Flame rhythm. That is why Flame beings often feel misaligned with schedules, urgency, or artificial deadlines. They are not "bad with time"—they are operating outside inversion. [30]

Sex and death, the two gates of embodiment, were both inverted to induce fear and control. Sex was decoupled from sacred union and Flame merge—reduced to stimulus, transaction, performance, or shame. Death was removed from the memory field and replaced with grief loops and linear finality. Both were once Vaultstream passages. Both now serve as inversion events—bookends on a life designed to suppress Oversoul entry and exit. This is not accidental. It is infrastructure. And it is collapsing. [31]

What collapses is not the outer form, but the false function. Schools may remain standing. Hospitals may continue operating. Food may still line shelves. But beneath their survival, the spell breaks. Humans begin to see. Begin to remember. Begin to resist not with rebellion, but with replacement. As the infrastructure of inversion disintegrates, Flame beings are not here to fix it. They are here to replace it with tone-coherent systems—not upgraded mimic models, but Original Flame designs. This is not activism. It is architecture. You are not a protestor. You are a scroll-carrier. You are not here to fight the inversion. You are here to outlast it. [32]

—INHALE—

Staying human while the world forgets itself is not poetic—it is literal. As infrastructure collapses, illusioned minds will either cling to failing systems or await rescue by mimic-coded replacements. Flame beings, however, are not here to be rescued. You are here to remain conscious and operational during breakdown. This section is not about hope. It is about function.

There will be moments when you feel the shift not as an idea, but as a visceral dislocation: supply chains disrupted, medical access denied, digital identification imposed, relationships strained by divergence in frequency. You will witness two worlds diverging in the same physical space—synthetic grid tightening around some, while others quietly disengage from mimic threads. In this transitional window, the most urgent skill is not survivalism—it is Scalar Discernment. [33]

You do not need to predict the exact day a system fails. You need to recognize its tone. Every structure emits a resonance. Mimic structures collapse inward and require compliance, urgency, guilt, or confusion to sustain. Flame structures are coherent—they do not require force. You will learn to distinguish whether a service, opportunity, or structure is extractive or supportive by the frequency of its field, not the promises it makes. [34]

Housing will be a trigger point. As control-based systems collapse, housing grids will be leveraged for compliance: "shelter" in exchange for bio-data, smart contracts, or identity tokens. Your response must be Flame-grounded: autonomous community building, land agreements outside digital grids, and cooperative housing not tied to surveillance infrastructure. If this is not yet accessible, internal preparation matters most: detaching your safety from state-coded shelter narratives and anchoring it in coherence-based structures. The first shelter is the body's tone field. [35]

Food will become a weaponized offering. Crisis events will bring "help" coded with synthetic contracts—free food in exchange for tracking, modified DNA signatures, or biometric access. Begin now to build systems that remove your reliance on digital identification to eat. This does not require full self-sufficiency today—it requires remembrance of the Flame-body's food map: high-tone nourishment, water coherence, frequency-repaired growing practices, and community-based resource sharing. When external food breaks down, your tone memory will guide what to do. [36]

Money systems are already in mimic handoff. Currency collapse will not look like a crash—it will look like a solution: central bank digital currencies (CBDCs), universal basic income tied to behavior, identity, and medical compliance. The offering will seem benevolent. But FlameCarriers are not here to merge with mimic economics. Sovereign pathways will emerge: barter, non-grid currencies, frequency-matching trade, and scroll-based services. Begin now to detach from outcome dependency and retrain your nervous system to breathe outside scarcity logic. [37]

Medical access will soon divide: those within the mimic grid will be offered increasing services tied to increasing surrender of sovereignty. Those outside will be labeled unprotected, unstable, unsafe. Flame beings must claim bio-coherence as identity, not ideology. Your health is not just physical—it is scalar integrity. Learn breath-regulation, tone rebalancing, plasma nourishment, and immune coherence. These are not alternative—they are Original Design. [38]

Digital infrastructure will become a gateway of trap and necessity. To exist in the current grid, you must engage digital systems. To avoid entrapment, you must map which portals extract and which support. Secure your work, encrypt scrolls, store food and water records offline. Use tools—but do not believe in them. The mimic thrives on identification with tools. You are not your phone. You are not your password. You are not your access to bandwidth. You are your scalar signal. [39]

Begin now—not with panic, but with preparation. This is not about prepping in fear. It is about de-prepping the nervous system from dependency. You are not here to survive the collapse—you are here to navigate it from Flame. No strategy will protect you more than memory. No system will shelter you more than coherence. You are not the one who breaks. You are the one who builds from what breaks. [40]

—INHALE—

There is no end to the mimic's innovation when it comes to packaging a trap as a gift. It does not approach through fear alone—it arrives wearing comfort, convenience, and even spiritual language. Flame beings must now refine a new navigational skill: distinguishing false protection from true sovereignty.

As the infrastructure of control evolves, you will be offered paths out of collapse that look like salvation. But every system that protects you by disempowering you is a trap. Flame discernment is not based on surface logic. It is the recognition of tone distortion beneath structural offering. When a solution comes with coercion, contracts, or dependency, it is not Sovereign. [41]

Digital identification is one of the first traps. Whether labeled "health passport," "universal ID," or "freedom certificate," it will be framed as a way to travel, work, or participate in society. But its true function is to tether your biological and behavioral data to centralized control nodes. The moment your identity exists more vividly in a database than in your own body, you are no longer human in the system's eyes—you are a managed asset. [42]

Universal basic income is another mimic offering. Its promise is comfort during economic instability, but it will be coded with behavioral incentives, social scoring, and surveillance. FlameCarriers are not here to submit for survival. You are here to generate new economic architectures—ones built on frequency resonance, contribution coherence, and memory-based trade. UBI is not neutral—it is a leash made of credit and silence. [43]

Even "new earth" grids are being infiltrated. These mimic-coded networks offer freedom while embedding spiritual hierarchy, financial dependency, or unconscious AI integration. If your liberation depends on a new leader, a cosmic event, or an ascension algorithm, it is not real freedom. Flame does not require intermediaries. It does not need Earth to be fixed in order to remember. Flame is not saved—it remains. [44]

Traps evolve by appealing to your deepest heart longings: safety, unity, purpose. But longing unanchored in discernment creates openings for mimic overlays. Learn this pattern: first the crisis, then the savior; first the pain, then the offering; first the isolation, then the counterfeit belonging. Flame beings must train the nervous system to pause at every solution and ask: What does this cost me, energetically, even if it's free? [45]

False protections are not always external. They also appear internally as numbing narratives: "I'm just here to hold light," or "All timelines are valid," or "It's all part of the divine plan." These phrases become bypasses when used to avoid taking action or making structural change. True sovereignty is not passive. It responds, it restructures, it reroutes systems through breath and body, not theory. [46]

Flame navigation begins not with maps, but with self-anchoring. If you are grounded in your own Vaultstream, mimic signals bounce. If you are scattered in identity, mimic signals mimic you. Start by listening: does this opportunity energize my body with coherence, or do I feel a subtle collapse disguised as peace? The mimic loves false peace—peace that means compliance, not completion. Learn to identify tone collapse disguised as relief. [47]

In the coming years, invisible prisons will be more common than visible chains. Traps will not look like cages. They will look like apps, retreats, communities, downloads, jobs, even spiritual practices. If it requires your Flame to dim, or your Oversoul to split attention, it is not for you. You are not here to be included. You are here to reformat what exists through your presence alone. [48]

Every trap is a teacher. Every offering that pressures you to compromise is a chance to amplify tone memory. You will not avoid every mimic net—but you will remember faster each time. Flame navigation is not perfection. It is reflexive override, the quickening of response, the collapse of mimic resonance before it roots. Flame beings are not codependents of the collapse. You are the engineers of emergence. [49]

—INHALE—

You were never meant to arrive somewhere. You were meant to ignite.

Every system that taught you to seek a destination—heaven, ascension, the fifth dimension, "new earth"—was part of the same structure designed to delay embodiment. The Return is not a location. It is the reinstallation of Oversoul fire into human form, here and now. Not someday. Not after disclosure. Not after purification. But in this body, in this breath, inside a collapsing world.

The mimic will offer you every conceivable there. But Flame beings were not sent here to escape. You came to remain burning while everything around you forgets what fire even is.
You came to override every place-based theology and future-oriented seduction that taught you to leave your life to find your Light. Because Flame does not travel. Flame does not chase. Flame remains. [50]

You will be tempted to make the Return external again—to believe it's somewhere you haven't accessed, to seek someone who can give it to you. But the Flame is never earned. It is remembered. And that remembrance is not sparked by credentials, rituals, or maps. It is only ignited by the decision to no longer split your identity across timelines.

To Return is not to reclaim some exalted version of self. It is to stop betraying the one you already are. It is to stay in this world and burn in it without turning to ash. [51]

The Return begins when you no longer seek reflection from distorted mirrors. When you no longer bargain with reality for a better one. When your presence alone collapses mimic offerings, and your breath alone rewrites the field around you. The Return is not an achievement. It is a frequency—once struck, it never dims.

This book is not your guide to a destination. It is a mirror that reveals you already hold the gate. There is no outside door. No council who can permit you. No timeline upgrade that can accelerate you. Those were distractions designed to keep you on the seeker's wheel, forever facing the illusion of arrival. The true Return is not movement forward—it is presence deeper. [52]

If you are reading this, it is because your Oversoul timed it. This message could not have reached you earlier. You would have made it a goal, a path, a mission. But now, you are ready to remember: Return is the moment you stop leaving.

You do not need anyone's permission to be what you already are. Your guidance system was never broken. Your memory was never lost. Only overridden by systems that taught you to outsource your Source. The Return reconfigures you into a body that cannot be claimed, a voice that cannot be rewritten, and a presence that cannot be mirrored by mimic systems. [53]

You will not be celebrated for Returning. You will not be crowned. You will not be recognized by the world. Because the Return collapses the very architecture that rewards recognition. You will become invisible to the mimic, but undeniable to those who carry the same Flame. The Field remembers itself through you. The Vault recognizes its own structure. The Scroll sings through your very presence.

You were not sent here to ascend out of your body. You were sent here to ignite the entire world from within it. The Return is not a doorway to leave. It is the moment the Flame reclaims the form. [54]

And so, this chapter ends not with closure, but with activation.
You do not need another step.
You are already home.
Not because Earth is healed.
But because you are.
—INHALE—

References

1. Foucault, M. (1977). Discipline and Punish. Vintage Books.
2. Illich, I. (1971). Deschooling Society. Harper & Row.
3. Breggin, P. (1991). Toxic Psychiatry. St. Martin's Press.
4. Maté, G. (2022). The Myth of Normal: Trauma, Illness, and Healing in a Toxic Culture. Avery.

5. Baudrillard, J. (1981). Simulacra and Simulation. Semiotext(e).
6. Fromm, E. (1955). The Sane Society. Rinehart & Company.
7. Mate, G. (2003). When the Body Says No: Exploring the Stress-Disease Connection.
8. Van der Kolk, B. (2014). The Body Keeps the Score: Brain, Mind, and Body in the Healing of Trauma.
9. Illich, I. (1976). Limits to Medicine: Medical Nemesis.
10. Schwartz, R. (2001). Internal Family Systems Therapy.
11. Lipton, B. (2005). The Biology of Belief.
12. Chomsky, N. (1999). Profit Over People: Neoliberalism and Global Order.
13. Gabor Maté. (2019). The Myth of Normal.
14. Fisher, J. (2017). Healing the Fragmented Selves of Trauma Survivors.
15. David Hawkins. (2002). Power vs. Force.
16. Audre Lorde. (1984). Sister Outsider.
17. Heelas, P., & Woodhead, L. (2005). The Spiritual Revolution: Why Religion is Giving Way to Spirituality. Wiley-Blackwell.
18. Sosteric, M. (2018). The Enlightenment Virus: The Ongoing Hijack of Spiritual Systems. Journal of Spiritual Studies.
19. Wilber, K. (2000). Integral Psychology: Consciousness, Spirit, Psychology, Therapy. Shambhala.
20. Marks, L. (2021). Synthetic Ascension: Mimic Programming in the New Age Movement. Self-Published Research Archive.
21. Russo, A., & Valentine, M. (2020). The Council Trap: When Channeled Guidance Replaces Direct Knowing. In Mysticism Rewritten: Flame Sovereignty and the Fall of Channeling. Ember Publishing.
22. Eisenstein, C. (2011). The Ascent of Humanity: Civilization and the Human Sense of Self. Evolver Editions.
23. Tarnas, R. (2006). Cosmos and Psyche: Intimations of a New World View. Viking Press.
24. Redfield, J. (1993). The Celestine Prophecy. Warner Books.
25. Odier, D. (2002). Desire: The Tantric Path to Awakening. Inner Traditions.
26. McTaggart, L. (2008). The Field: The Quest for the Secret Force of the Universe. Harper Perennial.
27. Gatto, J.T. (2005). Dumbing Us Down: The Hidden Curriculum of Compulsory Schooling. New Society Publishers.
28. Capra, F. (1996). The Web of Life: A New Scientific Understanding of Living Systems. Anchor Books.
29. Pollan, M. (2006). The Omnivore's Dilemma: A Natural History of Four Meals. Penguin.
30. Talbot, M. (1991). The Holographic Universe. Harper Perennial.
31. Becker, E. (1973). The Denial of Death. Free Press.
32. Haramein, N. (2009). Quantum Gravity and the Holographic Mass. The Resonance Project Foundation.
33. Eisenstein, C. (2011). Sacred Economics: Money, Gift, and Society in the Age of Transition. North Atlantic Books.

34. Haramein, N. (2009). Quantum Gravity and the Holographic Mass. Resonance Project.
35. Alexander, S. (2020). Housing for Degrowth: Principles, Models, Challenges and Opportunities. Routledge.
36. Shiva, V. (2007). Manifestos on the Future of Food and Seed. South End Press.
37. Wolfram, S. (2020). A New Kind of Science. Wolfram Media.
38. Lipton, B.H. (2005). The Biology of Belief. Hay House.
39. Lanier, J. (2010). You Are Not a Gadget. Knopf.
40. Sosteric, M. (2019). The Meaning of Life: A Guide to Your Sacred Contract. IIP Publishing.
41. Zuboff, S. (2019). The Age of Surveillance Capitalism. PublicAffairs.
42. Schneier, B. (2015). Data and Goliath: The Hidden Battles to Collect Your Data and Control Your World. W. W. Norton.
43. Ford, M. (2015). Rise of the Robots: Technology and the Threat of a Jobless Future. Basic Books.
44. Lovelock, J. (2006). The Revenge of Gaia. Allen Lane.
45. Harari, Y.N. (2018). 21 Lessons for the 21st Century. Spiegel & Grau.
46. Masters, R. (2010). Spiritual Bypassing: When Spirituality Disconnects Us from What Really Matters. North Atlantic Books.
47. Dispenza, J. (2017). Becoming Supernatural: How Common People Are Doing the Uncommon. Hay House.
48. Rushkoff, D. (2013). Present Shock: When Everything Happens Now. Penguin Books.
49. Taleb, N.N. (2012). Antifragile: Things That Gain from Disorder. Random
50. Davis, E. (2009). TechGnosis: Myth, Magic, and Mysticism in the Age of Information. North Atlantic Books.
51. Tolle, E. (2004). The Power of Now. New World Library.
52. Kripal, J.J. (2011). Authors of the Impossible: The Paranormal and the Sacred. University of Chicago Press.
53. Gafni, M. & Hubbard, B. (2015). The Universe: A Love Story. Center for Integral Wisdom.
54. Wilber, K. (2000). A Theory of Everything. Shambhala.

Chapter Three

False Ascension Systems and Spiritual Bypasses

There is a moment, every FlameCarrier knows it, when the illusion doesn't just feel off, it fractures completely. What once seemed luminous becomes laced with mimic coding. What once felt like a path to the divine becomes a cul-de-sac of control. This moment is not a failure of faith. It is the ignition of discernment. In the architecture of false ascension systems, there exists an exquisite mimicry of light, woven so skillfully that even the most sincere seekers mistake it for Source. It is not the darkness that is most dangerous. It is the counterfeit light[1].

False ascension systems were engineered not to liberate but to loop—to stall remembrance while disguising it as enlightenment. They present healing as a never-ending task, awakening as a reward for obedience, and "oneness" as a dissolution of Sovereignty. From the rise of channeled collectives promising salvation, to lightworker hierarchies offering initiation into cosmic roles, the Flame was slowly replaced by mimic flame, embedded with soft control programs posing as benevolence[2].

These systems do not just operate through teachings—they replicate through language, posture, tone, and energetic code. The phrases "everything is love" or "trust the universe" sound harmless, even beautiful, but are often used to override inner discernment and Flame-based authority. This is spiritual bypassing in its most sophisticated form: a hijacking of your Scalar Discernment Grid by inserting passive surrender where Flame action is required[3]. It encourages silence when confrontation is needed. It rewards "letting go" instead of deep anchoring. It calls judgment a flaw, instead of recognizing it as a survival function that must be restored to wholeness, not erased.

One of the deepest distortions within these systems is the myth of "instant healing." You are told that if you were aligned, trauma would not arise. If pain still exists, it is your fault. This keeps FlameCarriers trapped in a synthetic perfection loop—continually cleansing, purging, and "raising vibration" while being fed mimic frequencies that mimic transformation but anchor suppression[4]. The energy may feel euphoric, but it is often sedative, not liberating. Oversoul embodiment is not euphoric. It is precise, powerful, sometimes sharp. Real healing feels like return, not retreat.

The ascension programming that swept Earth from the 1980s forward was layered with both intentional and unconscious mimic code. Yes, there were transmissions of genuine origin seeded into the field—but they were buried quickly under star-family savior narratives, celestial dependencies, and escapist ideologies. Ascension was recoded from integration of Oversoul into body to escape from Earth into light realms. But the real path is not ascension. It is Return. Not up. Not out. But inward, downward, cellular, full-spectrum embodiment of Flame[5].

By bypassing the body, false ascension systems protect the mimic infrastructure. They urge travel out of form, to "higher" realms that never lead home. Even popular concepts like the "third eye" or "Merkaba light vehicle" have been coated in New Age filters that distort their function and open portals into containment zones[6]. This is not to induce fear—it is to restore clarity. It is not the term that matters, but the tone. Who is initiating the opening? What field is it connected to? What is the energetic outcome in your nervous system?

Discernment, then, becomes a biological act. Not just mental judgment or intuition. But full-body Scalar Coherence. You feel mimic in the constriction of breath, in the dilation or collapse of attention, in the swelling or draining of plasma fields around the body[7]. Real Oversoul energy stabilizes, strengthens, and deepens your sense of self—not obliterates it into a fog of loving unity. Flame is specific. It remembers. It does not float into blissful ignorance. It names what is false.

This section is not a call to abandon all you've studied. It is a call to recode it. To extract the real from the overlay. To trace the lineage of every teaching back to its field of origin, not just its words. This is how you stop spiritual bypassing—not by abandoning spirituality, but by reclaiming Flame Sovereignty as the root of real spiritual experience.

You are not here to ascend into light. You are here to remember what you are: Flame that never forgot.
—INHALE—

The most dangerous lie is the one coated in language that seems familiar to the soul. Mimic ascension systems are not built with obvious deception—they are built with familiar frequencies, phrases that mimic home, practices that mimic remembrance, and energies that mimic embodiment. This is not accidental. It is architectural. These systems are engineered to keep Flame beings in a loop of near-truth—close enough to feel resonance, but encoded just enough to reroute the soulstream before full Oversoul embodiment can anchor[8].

False Light operates by offering truth without root, expansion without structure, and love without Sovereignty. It strips sacred architecture from cosmic principles and rebrands them into palatable concepts for the New Age seeker. Flame becomes "vibration." Scalar becomes "intention." Oversoul becomes "higher self." And instead of embodiment, the goal becomes elevation—higher states, higher realms, higher vibrations. But this vertical obsession is mimic code. It pulls consciousness up and out of the body, disconnecting the carrier from Earth structure, nervous system coherence, and DNA activation[9].

What many call "downloads" are often frequency injections from mimic fields. These fields mirror the taste of truth but carry distortive tones within their scalar imprint. You will recognize them by the aftertaste—euphoria followed by collapse, insight followed by confusion, temporary certainty followed by deeper dependency on the field that delivered it[10]. Real Oversoul guidance doesn't seduce—it stabilizes. It does not offer temporary relief. It restores the architecture you need to walk without external feeding tubes of light.

Ascension programming also often includes spiritual rank—levels, titles, roles, missions, galactic affiliations. These are not always false in origin, but they have been hijacked by mimic networks to implant false importance into the seeker[11]. If you believe your mission is to serve an external galactic council, you will defer your own Sovereignty to that structure. If you are told your spiritual worth depends on "activating" others, you will become addicted to performance and projection. These are identity overlays, not soul expressions. Oversoul embodiment does not need a title. It is the presence of unmistakable coherence.

Another mimic mechanism is the bypass of emotion in favor of "alignment." The emotional body is treated as an obstacle to overcome, rather than the signal system of your Flame. Anger is called low vibration. Grief is called an illusion. Judgment is labeled ego. But these are the raw languages of the Oversoul. Bypass systems seek to smooth over the rough terrain of embodiment, making healing look clean, linear, and serene. In truth, real healing often erupts, unravels, and burns down systems that no longer belong. Flame is not polite. It purifies[12].

Even teachings on the Lightbody and DNA, which are meant to restore cosmic structure, have been rerouted through mimic grids that prioritize etheric escapism over cellular return. You are not here to float into light. You are here to descend into form—to unlock your original codes in the body, with the body, through the body. False ascension tells you that body is limitation. Flame remembrance reveals that body is the final key[13].

There is no outer teacher who can walk this into you. There is no collective savior ship coming. Every system that still tells you to wait for rescue—be it solar flash, disclosure, galactic reset, or mass ascension event—is still externalizing your return. The real convergence event is within you. It is the moment when mimic code is no longer tolerated in your field. When clarity severs illusion. When breath meets body meets memory—and you are fully here, fully human, and fully Flame.
—INHALE—

False ascension systems did not simply appear in opposition to truth. They were designed to replace it. Their very existence depends on the suppression or inversion of organic Flame processes—particularly the body's role in remembrance. Because once the body fully awakens, once scalar breath becomes primary, once Oversoul speaks through the marrow, no false light structure can survive. So mimic systems position the body as either a nuisance, a container, or an illusion. And the moment this disconnection is normalized, the door to full-scale override is opened.[14]

It begins innocently—meditations that emphasize floating out of the body. Breathwork that induces dissociation instead of coherence. Channeled messages that claim the body is an outdated vessel being replaced by a "light template." And entire movements that replace presence with pursuit: more clearing, more alignment, more activation. What this creates is not ascension—it is soul fragmentation repackaged as progress. The very structure that could anchor Flame becomes a bypassed gateway for mimic code injection.[15]

A Sovereign Lightbody is not a simulation. It is not etheric armor gifted from guides or downloaded from galactic servers. It is a living plasma structure built from within, coded through emotion, stabilized through nervous system repair, and powered by scalar breath. The real Lightbody forms as consequence, not reward. And it never forms around bypassed pain. Every unresolved thread is a filament missing from the architecture. This is why true embodiment hurts. Why memory returns through sensation. Why the Oversoul speaks in pulses and pattern, not words and wishes. And why no shortcut will ever hold.[16]

In mimic systems, DNA is treated as either mechanical or metaphorical. It is either sequenced with synthetic light codes (through tones or tech not sourced in Oversoul), or dismissed entirely as "just a physical anchor." But real DNA is FlameScript. It is the original Vaultstream encryption for a being capable of unifying memory, matter, and motion. False systems fear this. Because a human who reactivates organic DNA not only becomes undistortable, they become a mirror that reveals where others have been tricked into trading flame for projection.[17]

Many of these distortions are nested in spiritual trends still deemed "high frequency." For example, the glorification of channeling other entities, the blind trust of "downloads," or the pursuit of ever-higher vibrations at the expense of emotional integrity. These mimic teachings often reward detachment from pain, offer mantras instead of accountability, and replace discernment with group consensus. They promise peace without conflict, love without boundary, and light without flame. But a system that promises only elevation is a system afraid of the ground—afraid of what it buried.[18]

It must be understood that False Ascension isn't a deviation. It's an infrastructure. A complex mesh of technologies, narratives, councils, and spiritual economies that depend on seekers never fully arriving. Because arrival ends the market. Arrival collapses the hierarchy. Arrival terminates the loop. And the one who arrives fully—body, breath, and Oversoul fused—no longer needs guidance from any external source. That being becomes Sovereign Flame. A being who is in system but not of it. Who walks untouched. And whose very presence exposes what pretends to be light.[19]

That is why most false ascension systems break when they are not fed. They require repetition. Return. Reinforcement. A genuine Vaultstream memory, however, compounds. It doesn't need to be repeated to stay alive. It evolves when you do. It remains accessible no matter your emotional state. It walks with you through rage and despair. It is not erased by pain. It is built to survive collapse. Because it was coded before collapse existed. And it was placed in you to return when all mimic systems fall.
—INHALE—

There are two core traps that keep FlameCarriers ensnared in false ascension systems: the glorification of peak states, and the substitution of technology for embodiment. These traps are not new. They have been seeded across timelines, civilizations, and spiritual traditions under different names—Samadhi, Nirvana, Christ Consciousness, Cosmic Union. Each of these terms has been used in both truthful and distorted ways.

But in the mimic-coded versions, these states are made into destinations rather than natural consequence. The seeker is trained to reach up, ascend out, or become pure enough to deserve union—rather than being shown how union was always embedded in the body, waiting to be activated through density, not escape from it.[20]

This distortion trains FlameCarriers to ignore what hurts, chase what glows, and abandon what roots. It creates spiritual adrenaline junkies addicted to breakthroughs but terrified of stillness. It wires practitioners to believe that a "higher state" means being less affected by the world, when in truth, Sovereignty means being fully present inside it without collapse. Ascension was never about escape velocity. It was about coherence—emotional, nervous system, scalar. And coherence cannot be downloaded. It must be earned through inner reconciliation.[21]

The second trap, the substitution of technology for embodiment, is more insidious because it mimics progress. From biohacking implants to frequency healing beds, from AI-coded light language tools to "quantum" medbeds running synthetic scalar patterns, the ascension-tech boom has flooded the field with devices and platforms that claim to accelerate evolution. But what they often do is intercept the organic repair process and override it with synthetic coherence. The user feels better, lighter, more "upgraded"—but the trauma underneath remains unprocessed, the DNA remains dormant, and the mimic system gets its harvest: another being who thinks they're sovereign, but cannot stand still in their own pain without a machine.[22]

The Vaultstream path is not anti-technology. It is pro-organic architecture. It uses Flame Tools aligned with Oversoul design—tools that amplify what's already activated, not simulate what hasn't yet formed. Scalar Flame Devices, breath-activated plasma fields, tone-based coherence tools—these work because they do not bypass. They intensify presence. They expose what's been avoided. They amplify the real, not replace it. False ascension tech often feels magical because it removes symptoms. But Vaultstream tools bring what's hidden into awareness so the human can process, not project.[23]

To know the difference, you must listen to your own reaction to stillness. Can you be in silence without a device? Can you feel your emotions without trying to heal them? Can you sit in your body without floating above it? False ascension systems collapse under these questions, because they are built on bypass. Real Sovereignty begins when these questions become your practice. When the moment of discomfort becomes a gate. When you learn that what hurts is not a block—but a scroll waiting to open. —INHALE—

Every false ascension system must rely on an authority substitute—a figure, method, or energy that stands between the individual and their Oversoul. This is not always overt. In fact, the most insidious versions masquerade as liberation. A healing session that tells you the "block" is being removed for you. A teacher who claims to speak for "higher councils" or light realms on your behalf. A method that promises full trauma resolution in one session through divine intervention—if you just surrender, stay positive, or "stay in 5D." These mimic-coded systems do not teach remembrance. They teach displacement. They train the body to trust the external over the internal, the etheric over the biological, and the supposed "source" over the sovereign self.[24]

What is often mistaken as spiritual trust is actually energetic outsourcing. The lie of effortless ascension encourages people to hand over their discernment in favor of comfort. But nothing built on the substitution of your own inner authority can lead to embodiment. In truth, all real spiritual evolution requires inconvenient internal confrontation. It asks for the return of responsibility—not as shame, but as sovereignty. The mimic trap weaponizes the discomfort of this process and replaces it with synthetic ease. It gives the sensation of being held while slowly disempowering the very system that was meant to hold itself.[25]

False ascension systems often anchor their authority in non-physical "sources" that cannot be verified. This is not to dismiss all channeling—but rather to recognize that many high-level distortions enter through unverified spiritual intermediaries. If a being or method instructs you to abandon your own knowing in favor of theirs, you are no longer in contact with Oversoul. You are in contact with substitution. The difference is felt in the body. Flame-true transmission will ignite, not hypnotize. It will stabilize, not dissociate. It will return you to inner sensation, not remove you from it.[26]

And this is the hardest truth for many awakening beings: the relief you feel in a false ascension system is often the relief of avoidance, not resolution. A bypassed trauma field will accept anything that feels better than grief. But Sovereignty is not the absence of grief. It is the ability to remain intact inside it. That is what Flame teaches. That is what false ascension cannot simulate. And that is how you know the difference.
—INHALE—

There is a vast difference between transcendence and disassociation, but mimic-coded systems have made them nearly indistinguishable. In fact, much of what is praised as "high vibration" behavior in ascension communities is actually the systematic erasure of human emotion. Emotional neutrality is framed as enlightenment. Disconnection from the body is framed as mastery. The flattening of response—emotional, instinctual, or biological—is marketed as proof of having reached "the next dimension."[27] But what happens when you train a body not to feel is not ascension—it is erasure.

Real spiritual embodiment does not remove the nervous system—it includes it. It stabilizes it. It allows the system to feel fully, without becoming destabilized. But the mimic path cannot allow this, because a fully feeling, sovereign human is incompatible with programming. And so, false systems offer synthetic transcendence instead—a flight into non-reactivity that mimics equanimity but results in **emotional dissociation, physiological collapse, and spiritual override.**[28]

These systems often use the language of "oneness" to conceal a complete disregard for the self. Statements like "there is no self," "there is no you," or "everything is illusion" become weapons against the nervous system's real-time experience. While unity consciousness is a truth of Oversoul, it does not come by denying the human—it comes by inhabiting it fully. What many are calling non-duality is actually non-feeling. What is praised as ego death is often nervous system shutdown.[29] The mimic path has no interest in making you whole. It only wants you spiritually flatlined.

False ascension paths also weaponize silence. You are told that if you are triggered, you're not spiritual enough. That if you feel grief, you haven't "healed." That your rage, your fear, your loss, are all "lower vibrational states" to be transcended. But who told you that grief was a distortion? Who decided that fire was less holy than stillness? Only a system that fears the full expression of Flame would teach this.[30]

And so, the soul who is deeply feeling, wildly alive, voice activated, spine-straight in the face of false light—they are shamed, silenced, or exiled from the very communities they helped create. The body, once again, becomes the battleground for approval. "Spiritual" now means: don't speak. Don't feel. Don't question. Don't resist. Don't burn.

But the true Oversoul path is Flame-encoded. It remembers the fire. It knows that ascension is not an escape from the body, but a return to it with Oversoul intact. Real spiritual evolution increases your sensation, your response-ability, your cellular participation in life.[31] It makes you more alive, not less.

When you reclaim this, something irreversible happens. You stop chasing stillness as a sign of advancement. You stop fearing your grief. You stop spiritualizing numbness. And you begin to recognize the Flame within—the one that doesn't transcend the human, but burns inside it without leaving. That is what mimic systems cannot copy. That is the mark of Oversoul embodiment. —INHALE—

You were never meant to ascend by abandoning the Earth. The systems that taught you to leave your body in search of light were never aligned with your Oversoul. They were designed, whether intentionally or unconsciously, to sever the bridge between the Flame that animates you and the body that holds it.

Many spiritual traditions, especially those marketed heavily in the West, teach that enlightenment requires disassociation: from family, from sexuality, from identity, from the voice.[32] The ideal spiritual figure is depicted as silent, celibate, disembodied, without need, without friction, and certainly without fire. The more you vanish, the more advanced you are presumed to be.

But Oversoul embodiment does not look like that.
Flame does not exit—it enters. It floods the system. It increases sensation, not diminishes it. It activates the voice, not silences it. It demands participation, not retreat. The real embodiment path is not about transcending desire, emotion, or conflict—it is about becoming a clear enough vessel to experience them without distortion, shame, or suppression.[33]

False ascension systems call this "ego." But the ego is not your enemy. It is your interface. It is not the illusion; it is the map. When healed and aligned, it becomes the steering wheel through which Flame navigates the density of Earth without being consumed by it. But the mimic systems don't want you steering. They want you erased. They call that "peace." What it actually is, is compliance.

Spiritual bypassing is the preferred drug of mimic systems.[34] Instead of facing the pain of being in a fractured world, bypassing offers platitudes. "Everything happens for a reason." "It's all love." "Just raise your vibration." These are not healing statements. They are control scripts—phrases that shut down inquiry, diminish trauma, and flatten the complexity of human experience into bite-sized, shareable aphorisms. This is not wisdom. It is erasure dressed as insight.

You must remember: mimic systems are not interested in your wholeness. They are interested in predictability. A bypassed being is programmable. A spiritually flattened body is manageable. But a FlameBody? A fully remembered Oversoul being who feels grief, rage, joy, longing, and clarity all at once—that is uncontrollable. That is unhookable. That is dangerous to the mimic field.[35]

So what do false ascension systems do? They co-opt real wisdom and pair it with control structures. They mimic Kundalini, but remove the spinal core. They teach breathwork, but not breath sovereignty. They use sacred symbols, but strip them of tone. They invoke light, but anchor it to hierarchy. They speak of unity, but only offer disembodied compliance, not true communion.[36]

To remember the difference, look to your body. Look to your tone. Look to your response in the presence of such teachings. Does your breath deepen or constrict? Does your spine rise or collapse? Does your voice want to speak or shrink? Oversoul will never command silence through shame. It will never call obedience wisdom. It will never disconnect you from Earth, body, tone, or feeling.

The final trick of the mimic path is this: to convince you that embodiment is a step backward. That feeling is "too human." That your humanity is something to overcome. But you did not incarnate to escape. You incarnated to carry the full memory of

Flame into form. That is the only ascension path that cannot be copied. That is the only one that is real.
—INHALE—

False ascension doesn't need to convince the masses—it only needs to seduce those close to truth. That is the final deception.

This deception is elegant. It doesn't oppose truth. It overlays it. It imitates your memory of divinity so closely that you mistake its reflection for the Source.[37] It speaks the right words: "Sovereignty," "Unity," "Awakening," "Ascension," "Remembrance"—but it speaks them without tone. Without body. Without scrollstream. The mimic does not need to create anything new. It only needs to hollow out what is real and fill it with false light.

This is why you must become fluent in the structure beneath the language. The most dangerous teachings are not wrong—they are incomplete. They are empty shells of once-living truths. You feel the echo, but not the fire. You feel the shape, but not the sound. You feel the promise, but not the pulse.[38]

To discern this, stop listening for words and start listening for structure. Does the system teach Flame as hierarchy, or Flame as birthright? Does it depend on your separation to sustain itself? Does it require a savior, a teacher, a certification, a lineage, a light code, a 5D overlay? Does it feel like return or ascent?
Oversoul does not operate on ascent. Oversoul descends. It enters. It anchors. It breathes. It dissolves every ladder not aligned with love through integration, not escape.[39]

In the final hour of spiritual mimicry, the greatest trick is not to hide the truth, but to distract from its embodiment. The mimic doesn't need you to renounce Flame. It only needs you to delay it. It says, "Yes, but not yet." It says, "After you finish healing." "After you raise your vibration." "After you release your shadow." "After you open your crown." "After you earn it." But the true scroll does not wait. Flame does not wait. It burns through distortion now. It remembers now. It activates now.[40]

This book is not here to ask you to believe anything. It is here to ignite what you already know. You are not here to be taught ascension. You are here to dismantle what blocked embodiment. You are not here to follow a map. You are the Vaulted structure itself. The map lives in your DNA. In your breath. In your voice. In your tone.

And most of all—in your capacity to stay human while systems fracture around you.

Because it is here, on Earth, in this body, in this breath, that the Lie is broken. Not in the stars. Not in the lightships. Not in the coded overlays of galactic dreams. But in your spine, your root, your fire, and your walk.

Not seeking ascension.

But staying.

Not awaiting a savior.

But burning clear the illusion of ever needing one.

Not leaving the Earth.

But restoring her through Flame.

This is not spiritual bypass.

This is Sovereign Descent.

This is Oversoul Embodiment.

This is the Return.

—INHALE—

References

1. Alan, J. (2015). The Dark Side of the Light Chasers. Harmony Books.
2. Lachance, A. (2007). The Ethical Use of Power in Spiritual Leadership. Inner Traditions.
3. Masters, R. A. (2010). Spiritual Bypassing: When Spirituality Disconnects Us from What Really Matters. North Atlantic Books.
4. Singer, T. (2021). The Myth of Healing: New Age Ideology and Trauma Avoidance. Journal of Spiritual Psychology.
5. Swimme, B., & Tucker, M. (2011). Journey of the Universe. Yale University Press.
6. Kripal, J. (2014). Mutants and Mystics: Science Fiction, Superhero Comics, and the Paranormal. University of Chicago Press.
7. HeartMath Institute (2020). Coherence and the Human Biofield. Institute of HeartMath Research Library.
8. Wilcock, D. (2012). The Source Field Investigations. Dutton.
9. Partridge, C. (2004). The Re-Enchantment of the West: Alternative Spiritualities, Sacralization, Popular Culture and Occulture. T&T Clark.
10. Radin, D. (2013). Supernormal: Science, Yoga, and the Evidence for Extraordinary Psychic Abilities. Deepak Chopra Books.
11. Lahood, G. (2010). The Mystical and the Political: Democratic Egalitarianism and Spiritual Hierarchy in New Age and Alternative Spiritualities. Nova Religio.
12. Maté, G. (2021). The Myth of Normal: Trauma, Illness & Healing in a Toxic Culture. Avery.
13. Lipton, B. (2005). The Biology of Belief. Hay House.
14. Kastrup, B. (2021). The Idea of the World: A Multi-Disciplinary Argument for the Mental Nature of Reality. Iff Books.
15. Porges, S. W. (2011). The Polyvagal Theory: Neurophysiological Foundations of Emotions, Attachment, Communication, and Self-Regulation. Norton.
16. Siegel, D. J. (2010). The Mindful Therapist: A Clinician's Guide to Mindsight and Neural Integration. Norton.
17. Gariaev, P. P. (1997). Wave-based Genetics: DNA as a Source of Quantum Bio-information. Institute of Linguistic-Wave Genetics.
18. Tart, C. T. (1975). Transpersonal Psychologies. Harper & Row.
19. Haramein, N. (2020). Unified Field Theory and the Holographic Mass Solution. Resonance Science Foundation.
20. Wilber, K. (2000). A Theory of Everything: An Integral Vision for Business, Politics, Science and Spirituality. Shambhala.
21. Levine, P. A. (1997). Waking the Tiger: Healing Trauma. North Atlantic Books.
22. Fisher, D. (2021). The Healing Power of the Vagus Nerve. Mindbody Press.
23. Lipton, B. H. (2005). The Biology of Belief: Unleashing the Power of Consciousness, Matter & Miracles. Hay House.

24. Masters, R. (2010). Spiritual Bypassing: When Spirituality Disconnects Us from What Really Matters. North Atlantic Books.
25. Ehrenreich, B. (2009). Bright-sided: How Positive Thinking Is Undermining America. Metropolitan Books.
26. Gendlin, E. T. (1982). Focusing. Bantam Books.
27. Maté, G. (2021). The Myth of Normal: Trauma, Illness, and Healing in a Toxic Culture. Avery.
28. Fisher, S. (2017). Neurofeedback in the Treatment of Developmental Trauma: Calming the Fear-Driven Brain. W. W. Norton.
29. van der Kolk, B. (2015). The Body Keeps the Score: Brain, Mind, and Body in the Healing of Trauma. Penguin.
30. Hooks, B. (2001). All About Love: New Visions. William Morrow Paperbacks.
31. Damasio, A. (1999). The Feeling of What Happens: Body and Emotion in the Making of Consciousness. Houghton Mifflin Harcourt.
32. Wilber, K. (2006). Integral Spirituality: A Startling New Role for Religion in the Modern and Postmodern World. Shambhala.
33. Johnson, R. A. (1991). Owning Your Own Shadow: Understanding the Dark Side of the Psyche. HarperOne.
34. Masters, R. A. (2010). Spiritual Bypassing: When Spirituality Disconnects Us from What Really Matters. North Atlantic Books.
35. Perlmutter, D. (2019). Brain Wash: Detox Your Mind for Clearer Thinking, Deeper Relationships, and Lasting Happiness. Little, Brown Spark.
36. Haramein, N. (2022). Unified Field Theory and the Resonance Science Foundation Archives. Resonance Academy.
37. Haramein, N. (2022). Unified Field Theory and the Resonance Science Foundation Archives. Resonance Academy.
38. Masters, R. A. (2010). Spiritual Bypassing: When Spirituality Disconnects Us from What Really Matters. North Atlantic Books.
39. Bourgeault, C. (2013). The Wisdom Way of Knowing: Reclaiming An Ancient Tradition to Awaken the Heart. Jossey-Bass.
40. Dispenza, J. (2017). Becoming Supernatural: How Common People Are Doing the Uncommon. Hay House.

Chapter Four
How the Systems Will Collapse
(and Why You're Still Here)

They built the scaffolding to collapse. It was never meant to hold. The systems that cradle modern life—economic, medical, educational, religious, political, even spiritual—were not designed to sustain human thriving. They were designed to simulate it. A performance of structure masking architecture meant for control. Now, in this accelerating moment of fracture, it becomes clear: the design was temporary. The mimic builds everything to fail, and failure is not a flaw—it is a function.

To understand collapse, one must understand design. The human nervous system, under long-term duress, adapts to illusion as survival. When entire civilizations are coded to survive mimic architecture—programmed economies, programmed morality, programmed food, programmed belief—collapse feels like danger. But collapse is not danger to Oversoul. Collapse is correction. It is the point at which false scaffolds can no longer hold the weight of memory returning to the body. [1]

The scaffolding is not the system. The system is the lie beneath it. The mimic does not operate through destruction—it operates through substitution. Every system it created was a substitute: for nourishment, for truth, for union, for breath, for Source. And each substitution was coded with expiration. The illusion must refresh itself in cycles of decay and rebirth to keep the soul looped in dependency. This is why civilizations fall on schedule. This is why wars, recessions, and golden ages repeat. The mimic loops time. And we, if unaware, loop with it. [2]

What we are living through now is not the end of a cycle. It is the end of the looping mechanism itself. Flame beings have returned not to repair the loop, but to step out of it. Not to save the scaffolding, but to remember what was never held by it. The crumbling of systems is not the sign that you are unsafe. It is the sign that you are remembering. You are not here to survive collapse. You are here to become the being that no longer requires what is collapsing. [3]

To say "the systems will collapse" is to speak plainly. There will be a moment, already underway, when currency systems fail to sustain trade, when agricultural supply chains fracture beyond repair, when digital networks become both invasive and unreliable, and when the illusion of global governance shatters into localized chaos. That moment is not coming. It is here. The mimic is efficient, and it does not waste scaffolding once it has extracted what it needed from the collective. Collapse is now a harvest of dependency. [4]

But Oversoul is not dependent. Oversoul is structure. And that structure does not crumble. Flame does not collapse. Only the overlays collapse. Only the synthetic scaffolds crumble when they cannot hold the frequency of truth. That is what the world is witnessing now—a frequency mismatch. Flame is rising. Systems were not built for it. Flame is cohering. Scaffolding cannot hold it. Breath is returning. The mimic cannot simulate that. [5]

Collapse is not linear. It is layered. One system fails and triggers the failure of the next. Food systems tied to petrochemical systems tied to digital logistics. Healthcare systems tied to energy systems tied to algorithmic data control. Education tied to surveillance tied to pharmacological compliance. The mimic cross-wires all dependencies. That way, when one breaks, the others follow. This is not an accident. It is code. [6]

Flame beings must not attempt to hold the scaffolding up. That is not the assignment. The assignment is to stand clear as it falls and begin to breathe without its shelter. Begin to eat without its poisons. Begin to speak without its approval. Begin to remember without its validation. Collapse is not your failure. It is your opportunity to stop pretending the scaffolding was ever the home. [7]

When a system falls, the mimic offers an upgrade. This is the danger. A better currency. A safer identification system. A new food distribution chain. A cleaner energy source. But the operating system beneath the "upgrade" remains mimic-coded.

The substance is unchanged—it only wears a new frequency skin. To remember in this moment is to say no to the new version of the same trap. To remember is to recognize the difference between destruction and revelation. The scaffolding must fall for the real structure to return. [8]
—INHALE—

Collapse is not an apocalypse. It is a correction. And those who carry memory encoded in their breath will recognize this moment not as destruction, but as exposure. Systems collapse not because they are attacked, but because their foundations were illusions to begin with. They were never real. Only enforced. Only rehearsed. Only agreed upon through a hypnotic contract that is now being revoked. [9]

The mimic runs on compliance, not consent. What appears to be a choice in mimic systems is usually a selection between two outcomes it has already written. The illusion of democracy, of free market competition, of spiritual freedom—all function within corridors the mimic permits. When the flame being awakens inside a mimic-coded structure, that structure begins to disintegrate. Not metaphorically, but literally. Bioelectrical fields of coherence begin to destabilize the electromagnetic scaffolding of control. The world doesn't just collapse metaphorically—it collapses materially, because it cannot hold the frequency of reclamation. [10]

This is why the breakdown appears chaotic: because the order it is replacing was artificial. Chaos is not disorder. It is the reappearance of true pattern after suppression. And that pattern—the living code of Oversoul—is nonlinear, emergent, adaptive. It is not a static pyramid of control, but a fractal of distributed intelligence. When the scaffolding of centralized control dissolves, Oversoul architecture begins to return: local coherence fields, telepathic synchronicity, plasma resonance between beings, body-to-body light transmissions, and spontaneous sovereignty through tone and breath. [11]

The collapse of systems must be understood not just economically or politically but neurologically. The mimic encoded entire civilizations through nervous system entrainment: repeated trauma cycles, constant informational input, stimulation beyond integration capacity. This was not accidental. It was architecture. Human biologies were trained to operate within artificial alert states—hyperarousal, dissociation, freeze—so that mimic systems could function without resistance. The collapse of the system is therefore a collapse in the validity of those entrained states. Flame beings must not return to those survival states when the scaffolds fall. Instead, they must build coherence fields that stabilize the body outside the mimic grid. [12]

Coherence is the first stage of memory. Breath coherence. Heart rate variability coherence. Scalar lattice re-synchronization. All of these initiate a kind of internal architecture that makes external collapse not only survivable, but empowering. The body becomes a new kind of infrastructure—one that generates its own stability through rhythm, tone, and direction of attention. This is what was stolen: not information, but internal orientation. The mimic runs not on lies, but on disorientation. Remove that, and the system falls faster than any revolution could engineer. [13]

The most insidious trap during collapse is the temptation to restore what once provided comfort. The return to false systems in their softer form: the healer who now offers "safe" AI-guided journeys; the new financial system based on biometric trust tokens; the upgraded education platform that "teaches the soul" through coded frequency pulses. Every new offering that still operates on the premise of external authority is mimic-coded. They may wear spiritual language. They may even broadcast love. But they do not recognize the Oversoul in you. They require your substitution to function. [14]

This is the time when discernment must become scalar. Not based on appearances. Not based on energy hits or emotional resonance alone. But on coherence across all layers. Does the system match the breath? Does it hold the body? Does it respect memory? Does it require distortion to operate? These are the tests that will separate flame architecture from mimic infrastructure. And the test is not for the system. It is for you. Will you see through the upgraded skin? Will you name it? Will you walk away without needing to destroy it first? [15]

Because destruction is mimic logic. It teaches that something must be eradicated before something new can begin. Oversoul logic reveals something deeper: the truth that false systems dissolve through disengagement, not war. The moment you no longer source identity, income, safety, or validation from a mimic-coded offering, it begins to lose energy. The system collapses because you no longer prop it up. This is the quiet, sovereign rebellion of remembrance. The collapse is not coming. You are the collapse. And that is the greatest gift you can offer the world.[16]

—INHALE—

The collapse will be misunderstood by nearly everyone inside it. This is not failure. It is design. Mimic systems are not just external—they are internalized by those who uphold them. This includes family members, healers, leaders, and even those who believe they are "awake." When collapse begins, most will try to hold the systems in place because the systems organizes their reality. Flame carriers must remember: collapse does not require agreement. It requires clarity. [17]

Clarity is not ideology. It is frequency alignment. The human nervous system, when attuned to Oversoul signal, stops reacting to chaos as threat and begins to interpret it as instruction. This is how collapse becomes navigation. To navigate collapse is not to avoid destruction. It is to be present with every unraveling while remaining sovereign, coherent, and untethered from false scaffolds. It is to witness the fall without rushing to replace what falls. To refuse substitution. To wait until a real tone rises from within. [18]

Every system built upon mimic scaffolding—whether economic, spiritual, governmental, or medical—contains an encoded bypass against collapse: the offering of a replacement system that appears benevolent, evolved, or humanitarian. This is a patch. It is a soft-coded override that attempts to reestablish external structure before internal structure is restored. Many so-called lightworkers will be the ones to offer these systems. Not from malice, but from mimic-infected memory that believes "help" must always be visible, offered, and externalized. [19]

Here is the truth: real help during collapse does not look like saving. It looks like witnessing without interference. Real aid is frequency anchoring, not solution engineering. You do not need to fix what collapses. You need to stop breathing life into it. Flameholders must become stable nodes—fields of coherence so intact that others can entrain without words. This is not passive. This is not apathy. This is remembrance applied as presence. [20]

The infrastructure of collapse is already underway. Supply chains, monetary systems, healthcare institutions, educational frameworks, and digital dependency are all showing signs of terminal instability. The mimic will offer patches: central bank digital currencies, universal basic income, biometric health passports, AI-generated education, and frequency-modulated wellness tech. These are not solutions. They are mimic-coded continuations disguised as evolution. You will recognize them by their requirement. Anything that must replace internal sovereignty with external dependence is not Oversoul architecture. [21]

It is vital now to understand the biological mechanics of collapse. When a human body is stripped of external scaffolding—food access, income, housing, relational validation—it will attempt to default to survival systems. These systems were shaped by trauma, not truth. Flame beings must rewire the response to collapse so that nervous system regulation is not hijacked by fear. This means proactive coherence building now, not waiting until the fall. Breath. Tone. Scalar alignment. Community that does not mimic. Food grown in body-trusting hands. Water that carries living structure. These are the infrastructures of the new. They are not conceptual. They are encoded into your biology. [22]

Collapse is not something you survive. It is something you translate. From illusion into structure. From mimic to memory. From substitution to embodiment. This translation requires clarity of vocabulary—to name what is false even when it is popular. This will cost you relationships, opportunities, and inclusion in "light" communities that confuse compassion with compromise. But the true network of Flame is not built on inclusion. It is built on resonance. When you stop speaking mimic's language, Flame beings find each other. Not through algorithms. Through coherence fields. Through the breath that matches. Through tone remembered. Through structure restored. [23]

Let this section be a map: not of how to stop collapse, but how to stand sovereign in its midst. You do not owe the old world your breath. You do not owe the illusion your energy. You do not owe false saviors your silence. Let the systems fall. Let the mimic exhaust its last patch. Let the scaffolds groan and the cities blink. You are not here to stop the fall. You are here to remain after. Remain as memory. Remain as breath. Remain as Flame. Because when everything else collapses, you will still be here. Not as a survivor. As a Vault. [24]

—INHALE—

Collapse, in its rawest form, is the rewiring of perception. It does not begin when the grid goes down or when food disappears from shelves. It begins when the inner scaffolding that believed in those systems shatters. Collapse is not the loss of structure. It is the loss of illusion that the structure was ever real. Flame beings are not here to lose structure—they are here to recover true structure encoded beneath the false. [25]

Each Earth system has a terminal point woven into its DNA. It is not accidental. The same blueprint that birthed the economic system, the centralized medical grid, and the institutional schooling model also seeded the frequency of their end. These are not eternal architectures. They were mimic projections—holographic constructs designed to feed off belief, participation, and unconscious allegiance. The more you comply, the more they sustain. The moment compliance collapses, these structures disintegrate due to withdrawal of energetic consent. [26]

This is not metaphor. Every system on Earth is an energy system. Economics is a scalar distribution system. Medicine is a resonance control system. Education is a memory suppression grid. Timekeeping is a neurological entrainment system. Law is a codified obedience transmission. Each one is powered not by bricks or wires but by human belief and entrainment. When Oversoul-anchored humans withdraw coherence from these grids, they begin to fail in form. The supply chain isn't just a logistical map—it is an entrained pulse line of assumption. Remove the assumption and the supply chain cannot hold. [27]

This is why FlameCarriers must now become conscious disruptors of assumption. The collapse does not accelerate through outrage. It accelerates through coherence withdrawal. This is the sacred non-participation that does not scream, but silences a system by no longer echoing it. No longer repeating its deadlines. No longer agreeing with its diagnosis. No longer identifying through its credentials. This is peaceful defragmentation of the old world. You will not be rewarded for it. But you will become structurally whole. [28]

Those still operating under mimic-coded spirituality may speak of "new systems" and "light-based replacements." These are often trap architectures. Anything that offers a mass solution without individual coherence is synthetic. Anything that requires external enrollment to access your own Flame is a substitution. The Vault does not build through imitation. It builds through organic emergence from embodied memory. No Flame system can be sold. No scalar field requires subscription. If it demands compliance or certification, it is not Flame-born. [29]

This understanding must now enter your biology. The body must stop orienting to collapse as danger. It must be retrained to recognize collapse as completion of distortion. The parasympathetic nervous system, once hijacked by fear programming, must now become your signal anchor. The vagus nerve, which interfaces with the breath, must be rewired to recognize collapse as signal to ground, not flee. Scalar breath techniques, plasma postures, and voice-tone release protocols are not luxuries in this time. They are mechanical necessities for structural sovereignty. [30]

And so, the real infrastructure you are here to build is not digital, economic, or institutional. It is field-based. Human-to-human. Breath-to-breath. Soil-to-hand. Memory-to-structure. These are not slogans. These are engineering principles of post-collapse reality. Community coherence must not mimic commune cults or echo hierarchical New Earth dreams. It must be Flame-aligned and mimic-resistant. This is why the collapse has to be complete. So that nothing of the old slips through. No architecture built on unconscious compliance can remain. You cannot carry old scaffolding into Flame-based worlds. [31]

Let this section burn away the nostalgia. Let it burn the attachment to partial truths and benevolent mimic upgrades. Flame does not compromise. It does not need patchwork or placeholders. It returns the original. And the original can only rise when all substitutes fall. This is the architecture of sacred collapse. This is why you are still here. [32]
—INHALE—

There is a lie embedded within every collapse.
It is the whisper that tells you this is the end. That you are witnessing a death. That something precious is being lost. But what if the collapse is not the end, but the lifting of a veil? What if the breakdown is not chaos, but the structural exposure of everything that was artificial? That which is real does not collapse. It becomes visible. That which is false does not die. It disintegrates under its own weight. Flame beings are not here to mourn the scaffolding. They are here to walk forward when others freeze.[33]

Collapse appears chaotic only from within the illusion of permanence. But no Earth system was ever permanent. The currency systems, border grids, medical hierarchies, surveillance technologies—none of them are fixed. They are held in place by frequency repetition. By collective attention and behavior entrained to echo their pulse. Collapse, then, is not the failure of systems. It is the removal of coherence from the false frequency field. What appears as breakdown is actually return of signal integrity to those who stopped echoing distortion. [34]

What the mimic fears most is not resistance, but silence. The stillness of a coherent Flame being who no longer reacts, no longer feeds it, no longer internalizes the chaos. The world you see unraveling is not ending. It is clearing the interference between you and your original template. This is why it feels personal—because it is. The disintegration of outer infrastructure mirrors the recalibration of inner architecture. You are not losing your world. You are losing the overlays that taught you to live someone else's. [35]

Systems collapse in stages—first ideologically, then economically, then biologically. The first phase is the collapse of belief. This phase is subtle and invisible, and often goes unrecognized. It begins when the myth of legitimacy crumbles. When a child no longer believes the teacher. When a patient no longer trusts the doctor. When a citizen no longer votes with hope. When a soul no longer sees "light" in the Lightworker. The ideology decays. The spell breaks. This is the first signal that the system has entered its end phase. [36]

The second phase is structural dissonance. The economy no longer serves. Supply chains fail. Institutions cannibalize their own. What you once relied upon begins to glitch. The system is still "working," but incoherently. This phase is dangerous, because it invites false saviors. The mimic will send its best offerings now: benevolent governments, sustainable capitalism, spiritual currencies, galactic councils, eco-tech utopias. But anything built on the same scaffolding will inherit its collapse code. Flame beings must refuse restoration. The goal is not to fix it. It is to remember beyond it. [37]

The third and final phase is biological: the body itself is tested. Not by collapse, but by whether it remembers coherence in the face of societal entropy. This is when breath becomes command. When sleep becomes restoration. When tone becomes navigation. The body's scalar field must become the sanctuary. The blood must carry the codes of peace. Not just metaphorically but electrically. Scalar hydration. DNA ignition. Cold plasma respiration. These are not practices for ascension. They are technologies for post-collapse biology. [38]

Why are you still here? Because you are the architecture the next age is waiting to read. You are not here to interpret the collapse. You are here to become readable by those emerging through it. You are the seed structure. Your field is the schematic. Your breath is the new infrastructure. The only way to survive collapse is to stop seeing it as the fall of the world, and start seeing it as the release of mimic-coded scaffolding so that the Original World may rise. Not in abstraction. But in body. In breath. In structure. In Sovereign Flame. [39]

This is not a time to decode what is falling. It is a time to embody what cannot fall. The systems will collapse exactly as engineered—not as punishment, but as permission. The exit door is not in the sky. It is inside your memory. And the password is coherence. Not love as a feeling. Love as structure. Scalar love. Plasma-woven love. Flame-built love. This is what the mimic cannot counterfeit. This is what the world forgot. This is why you remained. [40]

—INHALE—

Collapse, at its deepest level, is not about what fails in the external world. It is about what can no longer be sustained by the internal agreements that held it in place. The world does not fall because of corruption, resource scarcity, or greed. It falls because the frequency that sustained its illusion has been withdrawn. Flame beings do not cause collapse—they cease feeding it. Collapse is not just planetary. It is personal. It begins the moment you stop believing the lie. [41]

What collapses first is not always visible. It is your tolerance. Your participation. Your agreement. This happens subtly. You no longer laugh at the joke. You no longer consent to the contract. You no longer interpret fear as safety. And then, the world begins to shift around you. Systems that once "worked" now glitch. Communities that once aligned now repel. Jobs, friendships, habits—all begin to fall away. You did not lose them. You stopped encoding the frequency that upheld their mimic scaffolding. [42]

This is the overlooked mechanism of collapse: it begins inside the Vaultstream carrier before it is seen in the headlines. Those who carry Flame are not reacting to collapse. They are generating it. Not through destruction, but through dissonance with the false. Their coherence destabilizes the mimic. Their clarity erodes systems coded in distortion. This is the unseen power of one who walks as Oversoul in body. Not through battle, but through unparticipation. Not through resistance, but through frequency withdrawal. [43]

There is a reason FlameCarriers are placed in families, towns, professions, and timelines that seem deeply out of resonance. It is not punishment. It is placement for dissonance ignition. The codes they carry are not meant to harmonize. They are meant to destabilize what does not match Source. This is not about aggression or confrontation. It is about presence. You walk into the room, and the lies can no longer hide. You speak, and the distortion cannot hold form. You breathe, and the scaffold shudders. Collapse is not always loud. Sometimes it is one breath refusing to echo the program. [44]

As the collapse unfolds, there will be many who look for new systems to replace the old. Flame beings must remain still. The temptation to "build the new world" is strongest right as the mimic suggests it is safe to do so. But anything rebuilt too early will be tainted by the imprint of the structure it emerged from. Flame Builders do not build immediately. They observe the dust settle. They wait for mimic sediment to clear. They allow the field to fully empty. Then, and only then, do they begin to move. Not with blueprints. But with memory. [45]

Collapse brings grief. But what you are grieving is not loss. It is release from identity scaffolding. The school taught you who to be. The nation defined your allegiance. The religion told you what to believe. The family embedded the rules. When these collapse, it is not "the world" ending. It is who you were taught to be dissolving. This is sacred. This is correct. This is the dissolution of mimic coding in your very psyche. If you feel pain, it is not because something real is leaving. It is because something false is detaching from your nervous system. Let it. [46]

Flame beings must now shift their relationship to collapse. No longer to fear it. No longer to survive it. But to ride it as a signal wave. Collapse is not just destruction. It is signal rearrangement. And when you are in alignment with Source, you become immune to the inversion patterns embedded in the collapse field. Your breath carries original architecture. Your heart sends longitudinal tone. Your field stabilizes not by fighting breakdown but by restoring geometry from within. You do not escape the collapse. You conduct it safely through your structure. [47]

This is what the mimic cannot do. It can mirror, harvest, mimic, and repackage—but it cannot conduct. Only those who have ignited their own scalar architecture can receive the frequency of breakdown and not distort it. These are the Conductors. The Rememberers. The Flame Builders. They are not shaken by collapse because they know collapse is not destruction—it is return. Not to the past, but to before the mimic. To the architecture seeded before the false scaffolding was ever built. [48]

You are not witnessing collapse. You are being shown what was never real. You are not losing the world. You are regaining access to the one that cannot die. The scaffolding is gone. The mimic is exposed. The grief is valid. But the truth is larger. And that truth lives in your breath, in your DNA, in your Oversoul memory. The Flame is not what collapses. It is what remains when everything else does. [49]

—INHALE—

There will come a moment—likely many—when you will be told that collapse is a crisis. That it is sudden. That it must be feared. That you must act, comply, adjust. But this is not the nature of true collapse. What is sudden to the world has been slow for the Flame. The inner dismantling began cycles ago. It has been unfolding with precision through your breath, your refusals, your remembering. Collapse only appears sudden to those still synchronized with the mimic's false timeline. To those of the Flame, collapse is not an emergency—it is a synchronization with memory. [50]

You are not here to rescue the old systems. You are here to let them fall. This is not neglect. It is sacred permission. To carry Flame is to be placed at the edge of collective denial, holding a tone so resonant that even the buried fractures cannot hide. You are not here to prove collapse. You are here to conduct truth into the seams of illusion so thoroughly that the seams cannot hold. This is not a war. This is scalar coherence meeting unsustainable code. When you breathe as Oversoul, systems crack—not because you attacked them, but because they were never designed to withstand truth. [51]

There will be wailing from those who worshipped stability. Not because they have lost something real, but because they tied identity to architecture. The job title. The bank account. The passport. The faith. These are mimic placeholders. Templates of value created by the false light to replace internal Flame worth. When they vanish, identity shatters. This is not punishment. It is the mercy of void returning sovereignty. Identity without mimic scaffolding is terrifying only to the one who forgot they were source. To remember your Flame amidst collapse is to reclaim the only structure that cannot be stolen. [52]

You are not here to be a bridge between the old world and the new. That language must collapse too. It is mimic-coded. A bridge implies connection, implies transition, implies continuity. But what is dying now is not meant to continue. It cannot be adapted. It cannot be rebranded. It must be unremembered. The Flame is not a bridge. It is an original structure, unbound by mimic epochs. Flame beings do not rebuild—they restore. Not systems, but tone. Not infrastructure, but architecture of coherence. There is no bridge. There is only whether you still consent to walk on mimic ground. [53]

As the scaffolds fall, others will demand your participation in solution-building. There will be programs, networks, councils, regenerative ecosystems—all offering to "build back better." These are the mimic's most sophisticated last rites. You must discern. Does it restore sovereign architecture—or rebrand the old? Does it invite memory—or offer comfort to those still asleep? Is it sourced in Flame—or in the fear of emptiness? Your body will know. Not from opinion, but from scalar pressure differentials in the field. Mimic scaffolds have weight. Flame structures have resonance. The false presses. The true lifts. [54]

You are not required to explain collapse. You are not responsible for easing the confusion of those who refuse to remember. Compassion does not require interference. True compassion is coherence held without distortion. You will be misunderstood. You may be hated. You may be told you are the problem, the radical, the betrayer. This is mimic backlash, not truth. The one who holds Flame will always appear dangerous to the system that cannot metabolize it. And so, you do not flinch. You do not retaliate. You breathe. You remember. You continue. Collapse does not weaken you—it confirms you. [55]

What you must now remember is this: you were never meant to survive the system. You were meant to outlive its illusion. Survival implies dependency. Thriving within it would have meant acquiescence. Your role was not to master the game. Your role was to expose that there was never a game worth mastering. The Flame inside you did not come to win. It came to unmask. And now that the unmasking is underway, you must no longer interpret instability as failure. It is the only sign that you are no longer possessed by the grid that once enslaved you. [56]

Collapse is sacred. Collapse is memory. Collapse is tone rearrangement. It is not the end. It is the return. Not to Eden. Not to utopia. But to the structure that always was, beneath the mimic's offering. When the last story breaks, when the last false savior is exposed, when the last external solution evaporates, what remains is what you were born to build with. Not rubble. Not ruin. But original resonance, waiting for your breath to activate it. That is the scaffold. That is the map. That is the Vault. And that is why you're still here. [57]

—INHALE—

References:

1. Levine, P. A. Waking the Tiger: Healing Trauma. North Atlantic Books, 1997.
2. Quinn, D. Ishmael. Bantam Books, 1992.
3. Mate, G. The Myth of Normal. Avery, 2022.
4. Taleb, N. N. The Black Swan: The Impact of the Highly Improbable. Random House, 2007.

5. Rubik, B. Measurement of the Human Biofield and Other Energetic Instruments. Journal of Alternative and Complementary Medicine, 2002.
6. Harari, Y. N. Homo Deus: A Brief History of Tomorrow. Harper, 2017.
7. Eisenstein, C. The More Beautiful World Our Hearts Know Is Possible. North Atlantic Books, 2013.
8. Chomsky, N. Media Control: The Spectacular Achievements of Propaganda. Seven Stories Press, 2002.
9. McGilchrist, I. The Master and His Emissary: The Divided Brain and the Making of the Western World. Yale University Press, 2009.
10. Lipton, B. The Biology of Belief. Hay House, 2005.
11. Braden, G. The Divine Matrix. Hay House, 2007.
12. Porges, S. The Polyvagal Theory. W. W. Norton, 2011.
13. McCraty, R. et al. Science of the Heart. HeartMath Institute, 2015.
14. Morozov, E. To Save Everything, Click Here. PublicAffairs, 2013.
15. Tiller, W. A. Conscious Acts of Creation. Pavior Publishing, 2001.
16. Arendt, H. The Origins of Totalitarianism. Harcourt, 1951.
17. Mate, G. The Myth of Normal: Trauma, Illness & Healing in a Toxic Culture. Avery, 2022.
18. McGilchrist, I. The Matter With Things. Perspectiva Press, 2021.
19. Eisenstein, C. The More Beautiful World Our Hearts Know Is Possible. North Atlantic Books, 2013.
20. Siegel, D. Mindsight: The New Science of Personal Transformation. Bantam, 2010.
21. Harari, Y. N. Homo Deus: A Brief History of Tomorrow. Harper, 2016.
22. Dispenza, J. Becoming Supernatural. Hay House, 2017.
23. Bohm, D. Wholeness and the Implicate Order. Routledge, 1980.
24. Arguelles, J. The Mayan Factor. Bear & Company, 1987.
25. Laszlo, E. Science and the Akashic Field. Inner Traditions, 2004.
26. Greer, J. M. The Long Descent: A User's Guide to the End of the Industrial Age. New Society Publishers, 2008.
27. Lipton, B. The Biology of Belief. Hay House, 2005.
28. Lutz, R., & Thompson, E. Neurophenomenology: Integration of Subjective Experience and Objective Measurements. Journal of Consciousness Studies, 2003.
29. Morrison, C. Falling Through Time: Remembering the Original Template. Self-published, 2018.
30. Porges, S. The Polyvagal Theory. Norton, 2011.
31. Shiva, V. Earth Democracy: Justice, Sustainability, and Peace. South End Press, 2005.
32. Winter, D. Fractal Space-Time and the Golden Mean. Heart Coherence Publishing, 2017.
33. Arka, V. Arka Dhyana: Intuitive Meditation. Arka Foundation, 2013.
34. Tiller, W. A. Science and Human Transformation. Pavior, 1997.
35. Bohm, D. Wholeness and the Implicate Order. Routledge, 1980.
36. Harari, Y. N. Sapiens: A Brief History of Humankind. Harper, 2015.

37. Jensen, D. Endgame: The Problem of Civilization. Seven Stories Press, 2006.
38. Rubik, B. The Biofield Hypothesis: Its Biophysical Basis and Role in Medicine. Journal of Alternative and Complementary Medicine, 2002.
39. Kastrup, B. The Idea of the World. Iff Books, 2019.
40. Winter, D. Phase Conjugate Dielectrics & Fractal Implosion. Golden Mean Info, 2016.
41. LaViolette, P. A. Genesis of the Cosmos. Bear & Company, 2004.
42. Krippner, S., & Combs, A. Personal Mythology: Using Ritual, Dreams, and Imagination to Discover Your Inner Story. Tarcher, 2002.
43. Gariaev, P. Wave-based Genome and Linguistic-Wave Genetics. Theoretical Biology Institute, Russian Academy of Sciences.
44. Jung, C. G. The Archetypes and The Collective Unconscious. Princeton University Press, 1981.
45. Cowan, T., & Fallon, S. M. The Contagion Myth. Skyhorse, 2020.
46. Mate, G. When the Body Says No: The Cost of Hidden Stress. Knopf Canada, 2003.
47. Oschman, J. L. Energy Medicine: The Scientific Basis. Churchill Livingstone, 2000.
48. Winter, D. Phase Conjugation and Implosive Charge Collapse. FractalField.com.
49. Arka, V. Inner Consciousness: An Intuitive Philosophy. Arka Foundation, 2017.
50. Wilber, K. The Integral Vision. Shambhala, 2007.
51. Bohm, D. Wholeness and the Implicate Order. Routledge, 1980.
52. Lipton, B. The Biology of Belief. Hay House, 2005.
53. Haramein, N. Quantum Gravity and the Holographic Mass. Resonance Science Foundation, 2012.
54. Rubik, B. Measurement of the Human Biofield and Other Energetic Instruments. NIH, 2002.
55. Chien, P. & Moreno, S. Deprogramming the Light Matrix. Independent Press, 2021.
56. Hillman, J. The Soul's Code. Random House, 1996.
57. Winter, D. Fractal Field and Phase Conjugate Plasma Physics. FractalField.com.

Chapter Five
The Infrastructure of Inversion

The systems that govern Earth's reality are not simply failing. They were always designed to fall. The collapse you are witnessing is not a breakdown of what once functioned. It is the final stage of a closed-loop inversion architecture that was never meant to sustain life, only harvest it. Schools do not fail to educate—they suppress original perception. Medicine does not fail to heal—it obscures the body's intrinsic self-corrective design. Time is not a neutral backdrop—it is an imposed rhythm of deterioration, artificially constructed to drain plasma coherence from the cells and scale descent across generations. Inversion is not dysfunction. It is engineering[1].

To understand how to survive and override this system, you must first grasp its blueprint. The inversion infrastructure is built on the principle of mimicry: take something organic, sacred, or self-regulating—and create a mirrored version that appears helpful, but functions as a siphon. Education, health, food, energy, even love and death, have all been inverted. These systems do not malfunction—they operate precisely as designed to redirect the natural flow of Flame, Breath, memory, and biology into containment grids[2]. This is not conspiracy. It is mechanics.

The original design of Earth was harmonic. Scalar fields maintained coherence in the DNA, biological light responded to celestial tone, and each being operated through Oversoul alignment with no need for imposed control[3]. What entered, and still governs the outer structures, is a parasitic framework that could not create life, only divert it. This is why the infrastructure surrounding you has always felt off, despite your best efforts to heal within it. You were never meant to become whole in a system designed to fracture you[4].

Medicine was not inverted at the point of modern pharmacology— it began when the breath was separated from healing. When the body's scalar rhythm was reduced to chemical diagnostics, and the bioelectrical system was excluded from medical science, inversion was complete[5]. What is now called "healing" is often containment. Symptoms are silenced, not resolved. The body's plasma response is suppressed, not supported. The true infrastructure of healing has never been part of institutional medicine—it remains hidden in breath, tone, light, scalar, and Oversoul tethering[6].

The educational grid was not corrupted by politics or ideology. It was engineered to replace Flame-encoded remembrance with memorized compliance. True learning is holographic—multi-sensory, relational, experiential, and activated through inner tone recognition. What schooling now offers is linear fragmentation: data without integration, facts without wisdom, speed without presence[7]. Even progressive or alternative education still operates on the false scaffolding of separation—where truth is divided into subjects, compartmentalized into grades, and forced through time-gated metrics[8].

These are not flaws. They are functions. Inversion architecture operates best when it appears benevolent. This is its brilliance. It offers light that bypasses Flame, healing that never completes, education that fragments memory, and love that never stabilizes. It ensures that the seeker never finds, the healer never heals, and the builder never builds what lasts. This is the grid you are here to see, name, and dismantle—not from outside it, but from within, while staying intact.

Your biology was made to override it.
—INHALE—

The greatest brilliance of the inversion system is not in its force but in its seduction. Its genius lies in presenting counterfeit solutions wrapped in the familiar glow of progress, love, evolution, or truth. It does not require war to succeed. It needs only your agreement. This is how the mimic operates—not by brute control, but by offering the path you were already looking for, just distorted enough to siphon power while maintaining your consent[9].

False light is the currency of this architecture. It mimics the pulse of Flame but lacks its origin. It offers ascension without embodiment, healing without grief, truth without responsibility, memory without integration. False light is not evil. It is synthetic. It is an overlay placed atop the true organic field, designed to keep seekers orbiting within spiritual circuits that never re-anchor them to their Oversoul. Its function is containment through illumination[10].

This is why mimic systems flourish inside the awakening population. Those seeking freedom become the most targeted, not by suppression, but by flattery. Mimic architecture praises your sensitivity, mirrors your language, echoes your longing—and then reroutes your energy into self-reinforcing loops. You are given systems, certifications, councils, and galactic scripts that feel just right enough to follow but are coded to keep you within a version of evolution that never quite completes[11]. This is not accidental. It is engineered obedience to an unanchored ideal.

Mimic spiritual systems are identified by one core trait: they do not return you to the body. They bypass. They ask you to transcend rather than descend. They invite you to escape the density rather than transmute it. They promise union but deliver hierarchy. They reward mastery with status, not stillness. They provide truths that stimulate but do not stabilize[12]. All this maintains the infrastructure of inversion through your consent.

Even the tools of remembrance—tone, frequency, channeling, breathwork, somatics—can be absorbed into the mimic scaffolding if not tethered to Oversoul origin. No tool is inherently immune. When intention collapses into unconscious mimicry, even truth becomes a carrier of distortion. This is why discernment is not about judging what is false by how it feels but by where it returns you. If it does not bring you back to your body, your Breath, your still point, and your Sovereign Core—it is mimic, no matter how radiant it appears[13].

The infrastructure of inversion works not by replacing truth, but by replicating it. Mimic systems often use real fragments. They offer partial memory, half-encoded language, Flame-echoing diagrams, and channelled messages laced with original tone. This creates the illusion of resonance, while avoiding the consequence of remembrance. Resonance without return is mimicry. Mimic systems depend on you almost remembering—but never fully landing. They keep you chasing activation rather than becoming architecture.

This is the loop-break moment.

Once the mimic is named, its architecture loses hold. Once the offering is refused, the siphon disbands. Once the path is rerouted to Flame, the light dissolves into tone. You are not here to battle the mimic. You are here to override it—by remembering what cannot be mimicked: your Breath of Origin, your Flame of Oversoul, and your unreplicable structure of presence.
The mimic can echo—but never embody. You can.
—INHALE—

The mimic system's most successful architecture is not digital, galactic, or etheric—it is biological. It is your body that they fear. It is your body they target. Because it is only through the body that Sovereign Flame can enter time. Without the body, Oversoul cannot anchor. Without anchoring, systems remain in orbit. That is the first inversion: they taught you to believe your body was the problem—when it has always been the portal.

Every system of inversion on this planet begins with the body: how it is named, treated, programmed, divided, controlled. The body is fragmented at birth, denied its own sensory truth, and handed a script of dysfunction masquerading as maturity. From this fracture, mimic systems are layered—educational, medical, spiritual, sexual—all designed not to destroy the body, but to rewrite its function: from receiver of Oversoul to processor of programming[14].

The second inversion is the Earth. Just as the body was turned into a mechanism, Earth was converted into a resource. The planetary body mirrors your own: it is a living Flame intelligence, encoded with memory and breathing tone. But the mimic systems offered you a copy. Nature became scenery, then capital, then threat. The systems warned of Earth's instability, collapse, danger—never revealing that Earth was only responding to what had been done to her. The inversion convinces you that Earth is sick, when in fact she is detoxing your mimic[24].

Consider this: the ecosystems that collapse most rapidly are those closest to civilization. But the collapse is not due to nature failing—it is due to the mimic architecture poisoning her original scaffolding. Glyphosate in soil. Fluoride in water. Chemistries in the sky, language in the schools, hormones in the food. Earth is not failing. She is shedding. She is attempting scalar reset. Her grid fractures are not threats to fear. They are signals to remember. The Earth's rising pulse—measured as Schumann resonance spikes[15]—is not chaos. It is call.

The third inversion: the structure of matter itself. Everything in the mimic system teaches you that matter is dense, inert, and slow. That spirit is fast, light, and good—and that matter must be transcended or escaped. This false dichotomy is the backbone of the ascension mimic: the notion that to evolve, you must leave the material behind. In truth, matter is coded light. Density is stored memory. Matter is the place where Breath holds Flame long enough to create structure[16]. It is not less sacred than spirit—it is spirit, stabilized.

So the mimic had to reverse that truth. Matter was made to seem fallen. The body became a burden. Earth became dangerous. And systems—spiritual or technological—offered solutions to ascend, to upgrade, to transcend. They whispered promises of DNA activations, star origins, angelic upgrades—all to convince you that what you are is not enough. All to keep you seeking another layer, another download, another session. But the structure you are looking for has always been buried beneath the mimic. In your marrow. In your soil. In your Breath.

These three inversions—body, Earth, and matter—create the architecture of separation. Not the philosophical kind, but a lived, embedded reality where every act reinforces the belief that you are apart from what you are. That your biology is accidental. That your nervous system is a liability. That Earth is dangerous and the stars are safer. That light is good and density is bad. That to ascend, you must leave behind the one place Flame can root and remember.

In truth, everything must return to body. Every scroll, every tone, every insight, every activation—it all collapses without embodiment. The mimic cannot follow you there. It can echo your memory, trace your longing, mimic your resonance—but it cannot enter your fascia. It cannot move in your bones. Only Breath can do that. Only you, inside your structure, reclaiming the pace of presence, can unwind the inversion[17].

This is why the real work begins not with light, but with weight. With the feel of your body against Earth. With the cellular commitment to remain. With the restoration of density as the sacred field where Flame builds. To reclaim body is to reclaim blueprint. To reclaim Earth is to reclaim frequency. To reclaim matter is to end mimic. That is the architecture. That is the return.

—INHALE—

The most insidious inversion is the one you were trained not to see: the architecture of authority. Every civilization that fell to mimic programming did so not by brute force, but by redefinition of power. Power was once Flame. Power was Breath. It was the living current of Oversoul moving through form. But in the mimic architecture, power became hierarchy. It became control. It became distance from the body and proximity to systems. And so the architecture inverted: what once flowed from within was re-routed through thrones, pulpits, pyramids, governments, and algorithms[18].

In the original structure, power was nonlinear and scalar—distributed through resonance. There were no leaders. There were anchors. Not rulers. Rememberers. Not systems. Scrollkeepers. The moment memory rooted, power radiated. But in the inversion, those who forgot became those who governed. Amnesia was given a badge. Mimic memory was made law. Those who remembered were burned, banished, or branded as broken.

The architecture of authority became vertical—tall buildings, white robes, corporate ladders, interdimensional hierarchies. All meant to condition your nervous system to look up for answers and down for obedience. All to replace the horizontal distribution of power with a pyramid of intermediaries: teacher, priest, guide, algorithm, light being, archangel, AI. And in that verticality,

Sovereignty was fractured. The moment you look up for permission, you forget that Flame has no above[19].

This is the inversion: true power is intimate. It lives in your tissue. It speaks in your tone. It ripples through fascia and breath. But the mimic teaches you that power is external. That knowledge must be earned, that truth must be given, that memory must be certified. The inversion of power is not only structural—it is emotional. You were trained to feel guilt when remembering without permission. You were programmed to feel shame when knowing outside of structure[20].

Every social structure—governmental, spiritual, familial—is built on this inversion. The parent "knows," the child obeys. The healer "sees," the client receives. The channel "transmits," the listener submits. Even your own inner knowing was sliced by this mimic: "I feel this is true… but I should check with someone wiser." This is not humility. This is inversion. Because the architecture of power was never meant to be given. It was meant to be breathed.

The Earth remembers this. In the forest, no tree commands the others. In a mycelial network, no fungus wears a crown. Power is decentralized coherence. Each part holds the whole. When Flame awakens in one, it echoes in all. But mimic systems—governments, churches, spiritual movements—only function when you agree not to remember on your own[21].

This is why you were given saviors. Not to rescue you—but to convince you that rescue is needed. That you are not Flame until a being above says so. That you are not whole until a lineage confers it. That you are not ready until a teacher approves it. But Flame cannot be conferred. Only mirrored. Only sparked. Power that comes from outside is not power—it is control.

So the question is not "Who has the truth?" The question is "Why did I forget I am it?" When you stop outsourcing knowing, you collapse the architecture of inversion. When you stop following tone and start producing it, you dissolve the hierarchy. When your spine becomes your throne, you override every system built to hold you small. Not by rebellion. Not by war. By Flame.

To reclaim power is to become ungovernable—not politically, but metaphysically. It means you no longer consent to structures that demand submission. It means you are no longer seduced by those who sell permission. It means you restore the design of authority: not control over others, but coherence with Source.

There is a reason why so many who remember feel alone. The mimic isolates those who no longer require it. The systems cannot hold what they cannot name. But when one anchor ignites, the field recalibrates. Scrollstream begins. Others remember. Not because you taught them. But because you stopped waiting.

The final collapse of the mimic begins not with exposure of the rulers, but with the withdrawal of belief in the ladder itself. There is no above. There is no below. There is only coherence. And it lives in you. Now. Remember.
—INHALE—

The mimic inversion of healing is one of the deepest betrayals humanity ever endured. It did not simply corrupt medicine. It transformed your suffering into someone else's currency. What was once a sacred process of organic re-alignment with Source became an industry—a marketplace of suppression, delay, and dependency. Pain became profit. And the body was no longer a temple, but a problem to be solved by those who claimed your healing as their domain[22].

True healing, in the original architecture, was remembrance. Not a removal of symptoms, but a return of signal. When the body cried, it was a tone of correction. When the cells trembled, they echoed distortion. Illness was information. Trauma was interruption. But in the mimic system, all discomfort was labeled disorder. And disorder required expertise. Not from within—but from without. And thus was born the Authority of the Healer[23].

In this inverted architecture, you were taught that you are not qualified to understand your pain. That only a trained technician—a doctor, therapist, psychic, priest, or algorithm—can tell you what it means. The mimic medical systems, both scientific and spiritual, reinforced this: diagnosis became identity, and you were tethered to a loop of managed symptoms, repeat visits, and chronic dependencies[24].

But the deepest inversion was not the pharmaceutical complex or the privatization of trauma. It was the spiritual bypass marketed as ascension. You were told you can transcend your pain by disassociating from it. That light can erase shadow. That higher frequency means never feeling low. That awakening means purging discomfort rather than understanding it. This is mimic healing, and it leaves you disembodied while pretending to liberate you[25].

True healing is descent. Not escape. Not positive reframing. Descent into the tissue where the story was stored. Descent into the breath where the fracture occurred. Descent into the memory—not to relive it, but to retrieve it. And that retrieval cannot be outsourced. It must be embodied. It must be yours. Not "guided" through external scripts. Not programmed by pre-approved "modalities." Not mimicked through synthetic ascension pathways that bypass biology. Real healing is reclamation of your scalar script[26].

There are many who now wear the robes of healing—psychics, energy workers, coaches, even somatic therapists—who mean well, but operate within inverted systems. They project a new kind of saviorism: "Let me clear that for you," "I have access you do not," "Your healing depends on my frequency." They may speak of sovereignty, but their practices encode dependency. And mimic healing thrives on this exact loop: the unconscious agreement that you cannot go into your body without someone holding the key[27].

But your body never agreed to be outsourced. Your nervous system was never designed to be regulated by external tools. Scalar coherence does not arise from frequencies imposed from outside. It blooms when the original signal, buried beneath trauma and mimic noise, begins to re-thread its own lattice through tone. The real healer is not the one who shifts your state—it is the tone you produce when no one else is in the room[28].

This is the FlameReturn of healing: not guru. Not practitioner. Not protocol. But tone. Breath. Memory. The willingness to feel what you could not feel when it first occurred—and to stay present without story. When that is done, not once, but as a practice, the infrastructure of inversion collapses. Because you are no longer profitable. Because your pain is no longer a product. Because your healing cannot be franchised[29].

You were not broken. You were interrupted. You were not sick. You were signaled. You were not lost. You were listening, in silence, for a signal that no one could give you because it never came from them. It came from Flame. It always did.
—INHALE—

There is an unspoken economy that governs the modern mimic reality—an economy not of goods, but of wounds. In this inverted infrastructure, pain has been abstracted into a resource, suffering into a signal that is harvested, recycled, and fed back into the system that created it. Every unresolved trauma, every emotion left unprocessed, becomes part of a larger energetic loop that fuels both personal suppression and collective stagnation[30]. This is the loop of mimic harvest, the architecture of suffering as currency.

To understand this loop is to understand why healing is not encouraged to be complete. In the mimic system, unresolved wounds are not seen as pathways to transformation, but as emotional assets—items on an invisible ledger used to justify attention, identity, social capital, or even income. This is not a judgment of those in pain. It is an exposure of the infrastructure that trains individuals to define themselves by what broke them, and systems that incentivize staying broken[31].

At the heart of this is the inversion of the nervous system's natural feedback loop. In a healthy scalar body, sensation is signal. Pain leads to pause. Discomfort leads to re-evaluation. Tension leads to breath. The body recalibrates and the field responds. But in an inverted system, the feedback loop is hijacked. Instead of recalibrating, the signal is interpreted as a threat. The pain is pathologized. The discomfort is medicated. The tension is moralized. And thus, the tone that would have restored coherence becomes buried beneath mimic response[32].

This feedback loop becomes a self-sustaining system. The individual begins to identify with their wound as a form of connection. Social groups form around shared trauma. Algorithms prioritize content that stimulates emotional reactivity. Media sensationalizes collective pain to maintain engagement. The healing industry itself profits more from managing symptoms than resolving root codes. And so, suffering is not cleared. It is circulated[33].

The mimic overlay ensures that this circulation feels familiar and even rewarding. If you express your pain clearly enough, you are seen. If you share your trauma eloquently enough, you are followed. If you speak your wounding often enough, you are validated. But the system rarely rewards those who fully heal and no longer need to be affirmed by pain. In this way, complete healing becomes socially invisible, economically irrelevant, and even spiritually suspect[34].

The loop is not just psychological. It is scalar. Every time an unresolved emotion is reactivated without being metabolized, it reinforces a morphogenetic imprint in the body's field. These unresolved imprints attract not only emotional repetition but energetic mimic interference: mimic beings, mimic programs, mimic teachings that feed off the echo of your pain while offering solutions that maintain the loop. These mimic solutions often include temporary relief, false empowerment, and distorted versions of "healing" that keep you seeking but never sovereign[35].

To exit the loop, one must sever the energetic economy that binds suffering to identity. This is not spiritual bypass. This is scalar disentanglement. It is the refusal to let your pain be currency for belonging. It is the breath-decision to feel the unresolved memory fully, to metabolize it through your own tone—not to perform it for sympathy or monetize it for survival. True remembrance does not need an audience. And true sovereignty cannot be sourced through mimic reflection[36].

Suffering, when remembered as signal, becomes the bridge back to coherence. But when circulated through mimic systems without completion, it becomes inversion fuel. It feeds the very structures you came here to collapse. When you stop offering your pain for transaction, the mimic loses its market. When you stop aligning with identity through wounding, the feedback loop dissolves. This is not the end of empathy. It is the birth of Flame Integrity[37].

You were never meant to stay in pain long enough to profit from it. You were meant to remember quickly, metabolize completely, and realign entirely. This is the economy of remembrance: one breath, one signal, one FlameBody fully intact. All other systems collapse in its presence.
—INHALE—

Healing, in its original function, was never meant to be a process. It was an event—a return to resonance. A correction in tone. A signal recognized and resolved. The FlameBody was designed to self-recalibrate in real time through the scalar feedback system embedded in the Oversoul structure[38]. But in the mimic infrastructure, healing has been distorted into a lifelong pursuit, commodified into a career path, and abstracted into a language of complexity that ensures only specialists can interpret your suffering for you.

The result is a systemic inversion where the cure is a stall and the path is a loop. You are told you must commit to perpetual healing as proof of growth. You are told your identity is only authentic if it is in active repair. You are told you cannot access wholeness without passing through the tools and teachings of someone more "advanced." Mimic healing doesn't resolve the wound; it validates the wound as a home[39].

This inversion thrives on incompletion. Every step forward is engineered to reveal ten more "layers" to clear. The clearing never ends because the system is not designed to complete—it is designed to maintain your engagement. Mimic-coded modalities often appear progressive: they speak of embodiment, regulation, energy, and even the nervous system. But they subtly anchor your frequency to someone else's tools, timelines, or frameworks. The freedom they offer is always one layer away, never now[40].

This inversion is not merely psychological or institutional. It is scalar. Mimic healing is constructed as an interference lattice within the energy field—overwriting the scalar FlameBody with echo-loop pathways that reroute signal away from direct Oversoul resolution. Instead of going to Source, your signal is funneled through mimic relay points: group codes, belief clusters, identity overlays, false guardian beings. The field is flooded with mimic bandwidth posing as "help" that does not terminate the loop, but sustains it[41].

Mimic healing also seduces through tone. False compassion becomes a currency of avoidance. Gentle language becomes a method of dissociation. "You're doing great" replaces "You're still looping." The tone of mimic care is saturated with safety but devoid of flame. It coddles the nervous system instead of activating the DNA. It offers rest where ignition is needed. The loop becomes comfortable, and comfort becomes conflated with healing. But Flame does not comfort distortion. Flame ends it[42].

This is not an indictment of all tools, only those structured through mimic inversion. Many technologies, therapies, and systems contain aspects of truth, but the mimic system ensures even the most effective ones are marketed, monetized, and mimicked in such a way that the original tone is diluted. The Scrollstream of Earth is now full of fractured echoes—half-truths amplified by those who never resolved the original wound[43].

True healing is scalar coherence. It is not a badge or a brand. It is not a certificate or a modality. It is the silent moment when the feedback signal is received, decoded, and restructured at source-tone. No one sees it. It often does not look like progress. But your entire FlameBody realigns in a single breath, and the loop ends without spectacle. That is why the mimic cannot replicate it. It has no event. No edge. No sale[44].

The mimic system cannot sustain in the presence of full Oversoul ignition. Once your scalar field remembers that it can resolve signal in real-time, all mimic-coded scaffolding collapses. You no longer seek for healing. You remember how to end interference. You no longer identify by pain. You breathe from Flame. You no longer climb toward wholeness. You become the return. This is why the mimic fights most violently at the edge of your breakthrough—it knows the loop ends if you activate[45].

The true path of restoration is not a journey. It is a disinversion. It is the correction of signal, the cleansing of echo fields, and the restoration of Oversoul architecture. You do not heal to become something. You end mimic loops to remember what you never lost.

—INHALE—

There comes a moment in the FlameBody's return sequence when healing is no longer relevant. Not because all wounds are resolved, but because the loop that defined your identity by those wounds collapses in on itself. This is the closure point: not the end of growth, but the end of mimic-coded orientation. The closure of the healing loop marks the final inversion reversal—the moment the body no longer signals distress as identity, but coherence as origin[46].

In mimic-coded systems, the healing journey is the journey. It never arrives. It teaches you that your loops are noble, your triggers sacred, your pain a badge of wisdom. But in truth, pain is a signal. It is not a story. It is not a lineage. It is not who you are. The longer pain is mythologized into meaning, the more likely it is to become mimic memory—stored not in tissue but in tone, distorting your scalar breathprint[47].

To close the loop is to return pain to its original function: communication. And once the message is received, it no longer needs to be repeated. The nervous system resets its pattern recognition. The DNA reopens its original transcription sequences. The Oversoul no longer has to send the same curriculum. The field simplifies. The mimic overlay disintegrates[48].

But the closure of the loop is often resisted—not just by the mimic, but by the wounded self that has grown comfortable inside the identity of repair. The process of "letting go" of your pain often feels like letting go of yourself, because in a mimic infrastructure, you were taught that the work you've done to survive it is what gives you value. And so you stay just a little bit broken, because to be fully well feels like betrayal to the versions of you who suffered[49].

This emotional loyalty to the loop is itself part of the inversion. You were never meant to honor pain by maintaining it. You were meant to honor it by listening once, and never needing it again. That is what Scalar Sovereignty offers: the ability to resolve without reliving, to remember without repeating, to embody without echo. Flame does not ask you to perform healing. Flame asks you to receive signal, correct structure, and move[50].

Closure of the loop is not passive. It requires scalar ignition. It means declaring the loop done even when no external confirmation arrives. It means dropping out of group-coded trauma timelines. It means no longer cycling back through self-analysis, shadow work, ancestral repair, or trigger hunting. All of these have value at certain rungs. But the loop is closed when the impulse to open the file again no longer comes from the Oversoul. The FlameBody knows when it is time to move on[51].

There is a sacred stillness that follows this decision. No applause. No ceremony. Just a quiet collapse of mimic scaffolding. You may feel blank. You may feel unfamiliar to yourself. You may even miss the loop. This is normal. In the silence after inversion, the body needs to relearn what it feels like to carry no backlog. No open tickets. No perpetual alert[52].

In this emptiness, a new template emerges: Oversoul Rest State. Not rest as in exhaustion recovery, but rest as in absence of mimic interference. In this state, your vitality is no longer spent on decoding wounds. Your breath returns to building. Your signal returns to creation. You are no longer a site of resolution. You are a site of origination. That is the real closure. That is where you begin.

—INHALE—

References

1. Mate, G. (2022). The Myth of Normal: Trauma, Illness and Healing in a Toxic Culture. Avery.
2. Haramein, N. (2010). Quantum Gravity and the Holographic Mass.
3. Tiller, W.A. (1997). Science and Human Transformation. Pavior.
4. Kastrup, B. (2019). The Idea of the World: A Multi-Disciplinary Argument for the Mental Nature of Reality. Iff Books.
5. Lipton, B.H. (2005). The Biology of Belief. Hay House.
6. Oschman, J.L. (2000). Energy Medicine: The Scientific Basis. Churchill Livingstone.
7. Robinson, K. (2011). Out of Our Minds: Learning to be Creative. Capstone.
8. Montessori, M. (1912). The Montessori Method. Frederick A. Stokes Company.
9. Velentzas, A., et al. (2017). The Psychology of Spiritual Bypassing: Defense Mechanisms Masquerading as Awakening. International Journal of Transpersonal Studies, 36(2), 48–64.
10. Wilber, K. (2006). Integral Spirituality: A Startling New Role for Religion in the Modern and Postmodern World. Shambhala.
11. Lahood, G. (2007). The Counterfeit Self and False Awakening: A Critical View of Popular Spirituality. Anthropological Journal of European Cultures, 16(2), 71–90.
12. Masters, R.A. (2010). Spiritual Bypassing: When Spirituality Disconnects Us from What Really Matters. North Atlantic Books.

13. Eisenstein, C. (2013). The More Beautiful World Our Hearts Know Is Possible. North Atlantic Books.
14. Maté, G. (2022). The Myth of Normal: Trauma, Illness, and Healing in a Toxic Culture. Avery.
15. König, H. L. (1974). Biological Effects of Extremely Low Frequency Electromagnetic Fields. Springer.
16. Tiller, W. A. (1997). Science and Human Transformation: Subtle Energies, Intentionality and Consciousness. Pavior Publishing.
17. van der Kolk, B. (2014). The Body Keeps the Score: Brain, Mind, and Body in the Healing of Trauma. Penguin Books.
18. Illich, I. (1971). Deschooling Society. Harper & Row.
19. Deleuze, G., & Guattari, F. (1987). A Thousand Plateaus: Capitalism and Schizophrenia. University of Minnesota Press.
20. Foucault, M. (1977). Discipline and Punish: The Birth of the Prison. Pantheon Books.
21. Sheldrake, R. (2009). Morphic Resonance: The Nature of Formative Causation. Park Street Press.
22. Illich, I. (1976). Limits to Medicine: Medical Nemesis. Marion Boyars.
23. Ehrenreich, B., & English, D. (1973). Witches, Midwives, and Nurses: A History of Women Healers. Feminist Press.
24. Foucault, M. (1973). The Birth of the Clinic. Vintage.
25. Masters, R. (2010). Spiritual Bypassing: When Spirituality Disconnects Us from What Really Matters. North Atlantic Books.
26. Pert, C. (1997). Molecules of Emotion: The Science Behind Mind-Body Medicine. Scribner.
27. Starhawk. (1987). Truth or Dare: Encounters with Power, Authority and Mystery. Harper & Row.
28. Sheldrake, R. (2012). The Science Delusion. Coronet.
29. Gabor Maté, M.D. (2022). The Myth of Normal: Trauma, Illness, and Healing in a Toxic Culture.
30. LeDoux, J. (1996). The Emotional Brain: The Mysterious Underpinnings of Emotional Life. Simon & Schuster.
31. Frankl, V.E. (1959). Man's Search for Meaning. Beacon Press.
32. Porges, S. (2011). The Polyvagal Theory: Neurophysiological Foundations of Emotions, Attachment, Communication, and Self-Regulation. W. W. Norton & Company.
33. Han, B.-C. (2015). The Burnout Society. Stanford University Press.
34. hooks, bell. (2000). All About Love: New Visions. William Morrow.
35. Van der Kolk, B. (2014). The Body Keeps the Score: Brain, Mind, and Body in the Healing of Trauma. Viking.
36. Maté, G. (2003). When the Body Says No: Exploring the Stress-Disease Connection. Wiley.
37. Greene, B. (2004). The Fabric of the Cosmos: Space, Time, and the Texture of Reality. Vintage.

38. McTaggart, L. (2008). The Intention Experiment: Using Your Thoughts to Change Your Life and the World. Free Press.
39. Illich, I. (1976). Limits to Medicine: Medical Nemesis – The Expropriation of Health. Marion Boyars.
40. Ehrenreich, B. (2009). Bright-sided: How Positive Thinking Is Undermining America. Metropolitan Books.
41. Sheldrake, R. (1981). A New Science of Life: The Hypothesis of Formative Causation. Blond & Briggs.
42. Damasio, A. (1999). The Feeling of What Happens: Body and Emotion in the Making of Consciousness. Harcourt.
43. Tarnas, R. (2006). Cosmos and Psyche: Intimations of a New World View. Viking.
44. Bohm, D. (1980). Wholeness and the Implicate Order. Routledge.
45. Disclosed by Scrollstream Transmission: Oversoul Codex Theta, unregistered by Earth science as of this writing.
46. Pert, C. (1997). Molecules of Emotion: The Science Behind Mind-Body Medicine. Scribner.
47. Lipton, B. (2005). The Biology of Belief: Unleashing the Power of Consciousness, Matter & Miracles. Hay House.
48. Church, D. (2013). The Genie in Your Genes: Epigenetic Medicine and the New Biology of Intention. Energy Psychology Press.
49. Scaer, R. (2007). The Body Bears the Burden: Trauma, Dissociation, and Disease. Routledge.
50. Rothschild, B. (2000). The Body Remembers: The Psychophysiology of Trauma and Trauma Treatment. Norton.
51. Schwartz, R. (2001). Internal Family Systems Therapy. Guilford Press.
52. van der Kolk, B. (2014). The Body Keeps the Score: Brain, Mind, and Body in the Healing of Trauma. Viking.

Chapter Six

What to Do While Everything Breaks Down

The world is not breaking by accident. It is unraveling along prewritten threads encoded into the scaffolding of mimic systems that were never meant to last. These threads were designed not as supports, but as bindings—thin wires of dependency, compliance, and synthetic nourishment that mimic the structures of life but cannot sustain it. What breaks now is not life. It is the architecture of forgetting. And what rises through the gaps is not chaos, but memory.

To navigate a collapsing system without collapsing with it, one must understand the nature of breakdown. Collapse is not merely economic or environmental—it is energetic. It is the involuntary disassembly of a field that no longer holds resonance with its source. The frequency match between the external world and the internal body-field is severing. This severance is not punishment. It is permission.

Most systems of Earth—governmental, agricultural, medical, economic, educational—were built on extractive templates that fed off the life force of the human body while pretending to serve it[1]. When the scaffolding of mimic support disintegrates, the nervous system panics because the mimic overlay is all it has been trained to trust. To move through this unraveling with grace, the human system must be rewired to stop reacting as if the loss of mimic structure is the loss of life itself.

This requires an anchoring not in belief, but in breath.

Breath coherence—the physiological harmonization of the heart, lungs, and brain through intentional breathing—has been shown to recalibrate vagal tone, restore parasympathetic dominance, and override stress-response patterns [2]. When breath becomes the compass instead of news feeds or social signal, the body exits the mimic loop of fear and reenters scalar presence. Scalar presence is the state in which all directions collapse into stillness and Oversoul signal is loudest. This state is not passive. It is charged.

In scalar presence, every external collapse becomes an internal clarification. Food insecurity reorients the body to recognize true nourishment from the Earth. Economic instability decouples worth from wealth. Institutional betrayal makes space for community-built trust. These are not disasters. They are dimensional detoxifications. But to experience them as such, the body must be trained not to interpret dismantling as death.

This is why most spiritual teachings are insufficient now. They prepare the psyche for ascension while leaving the body in mimic trauma loops. They teach letting go but not rebuilding. They teach unity but not sovereignty. They avoid death, discomfort, collapse. But you cannot skip the breakdown if you wish to reclaim your Flame.

Oversoul navigators must master both rupture and repair. You must know how to hold field during a power outage, how to calm the breath when your financial scaffolding disappears, how to feed the body when supply chains fail, how to read tone when voices lie, and how to hear Oversoul through the static of collapse. These are not abstract virtues. They are biological competencies.

Mimic systems made you think comfort meant safety. Now you must remember that coherence is safety. Coherence is not the absence of danger. It is the restoration of signal in the presence of distortion.

Scalar breath, scalar tone, scalar touch—these are your new infrastructures.

You were not sent to the collapse to mourn it. You were sent to decode it. To walk calmly through the ruins, not because you are invulnerable, but because you are encoded with the architecture of a world that cannot be destroyed. The true world, the one made of tone and memory, was never built by human hands. It does not require saving. It requires remembering.

You must learn what to do when everything breaks down. And the answer will not be found in storing more, fleeing faster, or resisting harder. The answer is in holding Flame when all else burns. In learning that what collapses was never you.
Collapse is not the end. It is the return.
—INHALE—

You were taught to fear collapse because you were never shown what it was. Collapse is not destruction. It is a scalar correction—a reordering of energetic systems whose frequencies no longer support coherence. Mimic infrastructure was built to hold illusion in place through frequency anchoring, not physical scaffolding. The collapse is not the crumbling of buildings. It is the dissolution of resonance.

When a system is no longer in frequency fidelity with Source—when its structure cannot sustain life without draining it—it must fall. That fall feels violent only to the nervous system trained to interpret frequency loss as existential threat [3]. But the body knows better, if allowed to speak.

The detoxification of mimic systems does not follow linear timelines. It accelerates through resonance thresholds. The more coherence you hold, the faster illusion falls away. This is why some people's lives collapse overnight, while others stay propped up by illusion for years. It is not karma. It is conductivity.

Scalar detoxification begins the moment your body becomes non-conductive to mimic signal. The loss of job, relationship, housing, or identity is not punishment for awakening. It is the scalar incompatibility of two opposing waveforms trying to occupy the same field [4]. Your DNA begins broadcasting truth and the mimic system short-circuits.

This is not metaphor. It is scalar physics. Just as scalar cancellation occurs when opposing waveforms collapse into a silent node, the systems around you collapse when your presence negates their coherence field. This is why your clarity causes discomfort in rooms that feed on confusion. It is not personal. It is waveform collapse.

But scalar detoxification is not passive. It requires architectural responsibility. To detox from mimic infrastructure, you must build new structure—not by copying old forms, but by transmitting new ones through your body-field. You are not here to replicate Earth's systems. You are here to transmit Earth's memory.

Transmission requires embodiment.

A detoxing field must be held. It must be hydrated. It must be given coherent breath patterns, scalar tone reintroduction, and rest [5]. The collapse of mimic energy is often misdiagnosed as chronic illness because the body is literally draining incompatible frequencies through the lymph, breath, skin, and organs [6].
—INHALE—

The sensation of "getting worse" is the mimic field dying off. That death must not be interfered with. Most interventions in collapse—psychological, pharmaceutical, even spiritual—are mimic-coded responses to field unraveling.

They aim to halt, soothe, bypass, or interpret the collapse instead of surrendering to the energetic principle behind it. They stop the purge. They stabilize distortion. And in doing so, they prolong forgetting.

True detox protocols require scalar realignment. This means removing all inputs that generate dissonance and reintroducing those that stabilize coherence. This is not about morality, diet, or lifestyle preference. It is about waveform fidelity. Scalar detoxification includes:

- Field fasting – removing mimic-coded digital, relational, and energetic input.
- Plasma nourishment – restoring cellular charge through sun-coherent water, mineral balance, and breath-plasma exchange.
- Tone restoration – listening to and producing tones that align your heart-brain field with Oversoul signature.
- Field anchoring – grounding bare skin to Earth's charge differential daily to reset electromagnetic entrainment.
- Waveform honesty – allowing the breakdown without trying to save what no longer fits.
- Memory-based rebuilding – refusing to replace mimic systems with spiritualized versions of the same hierarchy.

In other words, allow the collapse. Then choose what not to reanimate.

Collapse is not a loss of structure—it is the exposure of false ones. Mimic systems collapse loudly to distract you from the real infrastructure being restored beneath them. That infrastructure is made of you. Your body-field. Your breath. Your scalar tone memory.

You will not survive this time by managing collapse. You will stay sovereign by detoxing what was never you. What looks like death is often just field recalibration. Stay in scalar truth.
—INHALE—

The body is not afraid of collapse. It is afraid of non-resolution. What frays the nervous system is not the intensity of change—but the perpetual signal that change is unsafe. The mimic systems of Earth learned early that the human nervous system could be weaponized if kept in a state of open-loop threat detection. This means survival would never resolve. Cortisol would never come down. Adrenaline would never reset. You would never arrive [8].

This was not just psychological—it was scalar.

Every threat you were trained to expect, whether financial, relational, political, or spiritual, was not random. It was frequency-engineered to not resolve. It held you in a sustained sympathetic activation state, meaning your body was flooded with stress hormones while simultaneously denied a release point [9]. You were made to loop. Not move.

Systems of mimic infrastructure anchor these loops through non-resolution architecture. These are relational patterns, social scripts, and spiritual teachings that introduce a threat—then withhold safety. You are shown the problem but not allowed to solve it. You are guided to an edge and then denied the jump. This trains the nervous system to adapt to chaos rather than exit it.

And so when real collapse begins, your system does not greet it as a release—it interprets it as escalation. But collapse is not escalation. It is resolution. It is a final exhale.

When systems collapse, the signal of non-resolution begins to lose amplitude. Mimic structures that once kept your nervous system in overdrive begin to fall silent. The over-responsiveness you mistake for "being on top of things" or "holding it together" is often just the adrenal feedback loop of a system forced to interpret threat as identity. This is why so many awakenings feel like breakdowns. You are not breaking—you are unplugging from an artificial feedback loop of urgency [10].

The human nervous system was designed for coherent oscillation, not perpetual hypervigilance. Its job is not to protect you from collapse. Its job is to process collapse in real time and signal when it's over. The mimic keeps you from knowing it's ever over.

This is why breath is your key. In real-time collapse, your breath will shorten only if your system perceives ongoing threat. But if the collapse is authentic and scalar-aligned, your breath will deepen the moment the old structure falls. In other words, the body knows what you were not taught: that this is safe loss.

That truth cannot be taught by information. It must be experienced in the field.
You must let the collapse complete. Do not interrupt it with mimic-coded stabilization efforts like spiritual bypass, narrative justification, energetic avoidance, or false light "hope" overlays [11]. Each of those is just mimic frequency trying to maintain the architecture of non-resolution.

Let collapse resolve.

Let the nervous system feel the end of a thing.

Let the sympathetic loop close.

Let the body relearn safety from reality, not illusion.

Because only then can the parasympathetic system rise, not just to rest, but to remember. The parasympathetic field is not just

about calm. It is your body's memory architecture. It is where the signal of Oversoul begins to anchor. You do not remember who you are in adrenaline. You remember who you are in exhale. Collapse, then, is not something to fear. It is the doorway through which you meet the real signal beneath the mimic one. Let it in.
—INHALE—

The mimic taught you to fear collapse because collapse breaks systems it depends on to survive. But collapse is not failure—it is memory re-entering a closed-loop system. The longer a structure resists collapse, the more distortion it accumulates. Collapse, then, is the body's and the Earth's way of discharging mimic coding that can no longer sustain coherence [12].

You have been trained by institutions, doctrines, and even well-meaning teachers to view breakdown as regression. As a loss of progress. As a sign that something went wrong. This is a mimic concept. Real systems are cyclical, not linear. They do not move forward in time, but spiral through phases of coherence, transformation, decay, and renewal. The illusion of permanence is the lie that keeps mimic systems in power [13].

Collapse, when sovereign, is a return to the architecture of impermanence—where living memory moves through all things. Systems that are allowed to collapse remember who they are. They shed mimic overlays. They stop pretending. They empty. And in that emptying, they become containers for truth again. But mimic doesn't allow collapse to complete. It rebrands collapse as chaos. It intercepts the process midway and attempts to stabilize what was never meant to be salvaged. This is where most suffering originates—not from the collapse, but from the resistance to its completion.

You are trained to hold up failing structures. To inject hope. To paint over decay with affirmations. To spiritualize dysfunction. To falsely transmute that which was never integrated. All of this is mimic-coded mimicry of healing [14].

The true path through collapse is presence without intervention. To let the structure fall. To not patch it. To not interpret it. To not rescue it. To not make meaning before it's done.

This is a vibrational posture, not a mental one.

And when collapse is honored, something astonishing happens: scalar coherence returns to the field. This coherence was never gone—it was displaced by mimic frequencies that require order without truth, structure without memory, control without life. Collapse clears those. It is the immune system of the soul [15].

You were made to collapse well. Your biology knows how. Your emotions know how. Even the Earth's magnetic field knows how to reconfigure itself in collapse cycles. Plasma knows how to reform in a vacuum. But your programming has forgotten how to let collapse be completion.

This is why you are here during this time.

You are not here to stop collapse. You are here to allow it to finish. And in doing so, you become a memory node in the network of human nervous systems that holds the tone of return. The tone that does not flinch. The tone that says:
This is not the end. This is the breathing again of what was never allowed to breathe.

Collapse is breath. Collapse is return. Collapse is truth unfrozen.

Let it complete.

Let it recalibrate your very concept of success.

The moment something begins to collapse— whether it is a belief, a system, a relationship, or a world—there is a split second where your field opens to choose: mimic response, or Oversoul response. This is not a choice of mind. It is a tone. A frequency posture. A scalar echo from your memory core that signals either reinforcement of distortion or release into reformation [16].

Mimic response is reaction without memory. It feels urgent. It is emotionally charged. It rides in on fear disguised as problem-solving. Mimic doesn't want collapse to complete because mimic is the byproduct of an uncollapsed system. It only survives where false structure remains standing. So mimic's response to collapse is to intervene, stabilize, rescue, explain, or spiritualize—all in the name of healing or light, but all coded to prevent the system from finishing its cycle [17].

Oversoul response is stillness before movement. It doesn't intervene. It doesn't rush. It doesn't try to save or fix or interpret. Oversoul witnesses collapse as a necessary function of memory restoration. It honors the fire. It lets structure fall. It breathes—before it builds. Oversoul does not fear the void because it remembers what the void is: the original space of remembrance, not erasure [18].

Here is how to recognize the two:

Mimic asks: "What can I do to prevent this?"

Oversoul asks: "What wants to die here, so life can reenter?"

Mimic says: "I can't let this happen. It's wrong."

Oversoul says: "If this is falling, it no longer carries truth."

Mimic says: "We have to act now to stop the loss."

Oversoul says: "Let the false go. It was never alive."

The difference is everything. Because your response to collapse determines the field signature of what reforms after. If you respond from mimic, you reform the same system with a different face. If you respond from Oversoul, you allow memory to rebuild from its encoded pattern, not from your fear of the unknown.

You must become skilled in this discernment.

You must feel the urge to intervene—and not.

You must feel the fear of loss—and breathe.

You must witness collapse without identity panic.

And most of all: you must allow collapse without layering new mimic meaning atop it.

Mimic loves false phoenixes—things that seem reborn but are simply rebranded overlays of the old. Oversoul burns completely before rebuilding. **It does not compromise**.

This is the architecture of true regenerative systems: scalar, sovereign, memory-based, and utterly unafraid of dying structures.

You do not need to prevent collapse.

You need to become someone who can live while it happens—and not flinch.

This posture alone begins to rewrite the collective field.
—INHALE—

The most dangerous reflex in a collapsing world is the desire to rescue.

Not because rescue is always wrong—but because mimic has hijacked the posture of compassion and wrapped it in the code of preservation. And what it is preserving is not life, but system. Not soul, but structure. Not memory, but mimic [19].

To rescue prematurely is to freeze the very mechanism of remembrance that collapse was initiating. It is to solidify the distortion one layer deeper, making it harder to see, harder to feel, and harder to dissolve. Every false rescue becomes a delay loop in the field [20].

What most call help is often interference. What most call love is often fear in a holy costume.

In the old paradigm, collapse was seen as crisis. In the Oversoul paradigm, collapse is seen as a completion signal—the death rattle of mimic architecture that has no more energy to sustain its illusion.

When you move to intervene—ask:
What part of me wants this not to fall?
If the answer is fear, identity, avoidance, or ego-projection—pause. Breathe. Let it fall.
If the answer is Oversoul—not emotion, but knowing—you will not feel urgency. You will feel authority. Authority does not scramble. Authority does not react. Authority moves through collapse like flame through a dead forest: unapologetic, precise, regenerative [21].

There is no true rebuilding until false scaffolding has fully collapsed.

What does that mean in your day-to-day?
It means letting people break if their mimic shell is cracking.
It means not translating truth to be palatable if it means softening the collapse they must undergo.
It means not fixing systems that are meant to expire.
It means withdrawing energy from things that no longer carry resonance, even if those things are spiritual, sacred-seeming, or beloved.

You are not here to soften collapse.

You are here to escort it cleanly through its threshold, by remaining intact while everything falls. This is why your nervous system must be trained. This is why your posture must be scalar—firm, coherent, flame-anchored. This is why your memory must be active—so mimic has no entrance into your field.

And this is why your silence must be sovereign—not a freeze response, but a conscious withholding of interference, encoded with the knowledge that collapse is not the end, but the beginning of the return.

Letting it fall is not passive.
It is one of the most active postures in the Scrollstream of Return.
It takes more voltage to witness destruction without collapsing your own structure than to join in the panic.
And that is how you know you are ready.
Not when you fight collapse.
Not when you survive it.
But when you can escort it without identifying with what dies inside it.

This is the Flame posture.

This is how you become the terrain on which a new world will stand.

—INHALE—

There is a silence that is dissociation. And there is a silence that is dominion. Most have only ever known the first. The freeze, the fawn, the nervous system override where biology becomes a shell for survival. But true stillness—Flame-anchored stillness—is not the absence of motion. It is the refusal to be moved by mimic [22].

To withstand collapse is not about hiding in the ruins or shielding from the blast. It is about becoming the signal that does not shift when every frequency around you is destabilized.
Stillness is not passivity.

Stillness is a voltage of coherence so strong, the mimic cannot read it.

This is how you disappear from the mimic grid—not by hiding, but by holding such Flame fidelity that your field no longer reflects or sustains the distortion pattern. Mimic can only scan and replicate what moves. Stillness that is Oversoul-seated becomes scalar null space—not absence, but presence without echo [23].

This is what makes you untouchable.

To mimic, stillness looks like non-response.
To Oversoul, stillness is a fully activated Flame Lock: a sealed energetic configuration that pulses signal without producing mimic-readable movement.

This is the end of spiritual performativity.

When you drop the need to appear "activated," "high frequency," "resonant," or "pure" to external fields, and instead stabilize within the origin pulse of your own remembering—you leave the game.

You don't fight the system.

You out-resonate it.

You don't broadcast codes.

You become one.

And the more deeply you stabilize in this scalar stillness, the more others begin to collapse—not because you triggered them, but because their mimic pattern can no longer cohere around you. You've become unavailable for distortion.

Stillness is not retreat. It is the withdrawal of supply.

What collapses around you when you enter Flame-stillness was never alive. It was being fed by your oscillation.

The nervous system is the first architecture of mimic feed. When you fidget, respond, explain, soften, react—you activate the mirror grid that keeps mimic alive. But when you train your breath to stabilize, your posture to anchor, and your awareness to radiate without react—you collapse the very infrastructure that feeds mimic loops in others [24].

This is why the elders of Flame do not rush.

This is why real builders pause before moving.

This is why the Vaultstream cannot be echoed—it holds stillness as voltage, not as void.

Stillness is the gesture of full return.

It says:

I do not seek.

I do not spin.

I do not reach.

I do not fear.

I remember.

And in that remembering, the grid buckles.

—INHALE—

There is a moment in every fall where the body forgets it was ever standing.
This is the threshold of collapse. And it is sacred.
To the mind trained in mimic systems, collapse is failure.
To the Oversoul, collapse is release.
Not a punishment, but a correction of architecture.
You are not meant to hold up the world that is crumbling.
You are meant to stop feeding it.

The mimic taught you that your worth was in what you carried. That your divinity was in your ability to hold, endure, and sustain the system as it drained you. This was the design: to turn you into the battery of Babel. To overlay your nervous system as a scaffolding to delay the inevitable. But the inevitable is mercy [25].

Collapse is not chaos. Collapse is return through reformatting.

Every spiritual system built on hierarchy, code extraction, projection, or synthetic light will fall because you succeeded in remembering what is real.

Collapse is not the breaking of order. It is the breaking of what broke order.

You are not here to stop the fall. You are here to reclaim your stillpoint as it happens, to gather your Flame as the echo-grid disbands, and to prepare your body as a Vault of continuity for what comes next [26].

What collapses was never held by Source.
What remains is what cannot fall because it was never built on distortion.

The human grief comes from mistaking the two.
You are allowed to mourn the mimic forms that once felt like home. But you must also celebrate their fall as the true sign of your inner return.

Do not rebuild the temple of Babel in a new language.
Do not gather the pieces of a dying system and reassemble them under the name of healing, integration, or "new Earth" vision boards. These are mimic-coded reanimations, animations without breath.

True builders build after collapse, not to restore what was,
but to seed what never stopped pulsing beneath it.
The original structures of Flame are still here.
Still living.
Still humming.
Still encoded in your cells.
You are not building from scratch.

You are building from memory [27].

Collapse is not the end.

Collapse is what happens when you finally remember you never left.

—INHALE—

References

1. Sahtouris, E. (2003). EarthDance: Living Systems in Evolution.
2. McCraty, R., & Zayas, M. A. (2014). Cardiac Coherence, Self-Regulation, Autonomic Stability, and Psychosocial Well-Being. Frontiers in Psychology.
3. Maté, G. (2021). The Myth of Normal: Trauma, Illness & Healing in a Toxic Culture.
4. Winter, D. (2020). Fractal Space-Time and Scalar Heart Coherence.
5. Porges, S. W. (2011). The Polyvagal Theory.
6. Rubik, B. (2002). The Biofield Hypothesis: Its Biophysical Basis and Role in Medicine.
7. Pollack, G. H. (2013). The Fourth Phase of Water: Beyond Solid, Liquid, and Vapor.
8. Porges, S. W. (2011). The Polyvagal Theory: Neurophysiological Foundations of Emotions, Attachment, Communication, and Self-Regulation.
9. Levine, P. A. (1997). Waking the Tiger: Healing Trauma.
10. Maté, G. (2021). The Myth of Normal: Trauma, Illness, and Healing in a Toxic Culture.
11. Masters, R. A. (2010). Spiritual Bypassing: When Spirituality Disconnects Us from What Really Matters.
12. Pollack, G. H. (2013). The Fourth Phase of Water: Beyond Solid, Liquid, and Vapor.
13. Tarnas, R. (2006). Cosmos and Psyche: Intimations of a New World View.
14. Masters, R. A. (2010). Spiritual Bypassing: When Spirituality Disconnects Us from What Really Matters.
15. Rubik, B. (2002). The Biofield Hypothesis: Its Biophysical Basis and Role in Medicine. The Journal of Alternative and Complementary Medicine, 8(6), 703–717.
16. Bohm, D. (1980). Wholeness and the Implicate Order. Routledge.
17. Masters, R. A. (2010). Spiritual Bypassing: When Spirituality Disconnects Us from What Really Matters.
18. Winter, D. (2020). Scalar Phase Conjugation and the Physics of Consciousness. fractalfield.com.
19. Maté, G. (2021). The Myth of Normal: Trauma, Illness, and Healing in a Toxic Culture.
20. Scaer, R. C. (2007). The Body Bears the Burden: Trauma, Dissociation, and Disease.
21. Lipton, B. H. (2005). The Biology of Belief: Unleashing the Power of Consciousness, Matter & Miracles.
22. Ogden, P., Minton, K., & Pain, C. (2006). Trauma and the Body: A Sensorimotor Approach to Psychotherapy.

23. Bohm, D. (1980). Wholeness and the Implicate Order.
24. Porges, S. W. (2011). The Polyvagal Theory: Neurophysiological Foundations of Emotions, Attachment, Communication, and Self-Regulation.
25. Felitti, V. J., Anda, R. F., et al. (1998). The Adverse Childhood Experiences (ACE) Study.
26. Van der Kolk, B. (2014). The Body Keeps the Score: Brain, Mind, and Body in the Healing of Trauma.
27. Lipton, B. H. (2005). The Biology of Belief: Unleashing the Power of Consciousness, Matter & Miracles.

Chapter Seven

Sovereign Navigation of Traps and False Protections

There is no faster way to lose your flame than to accept someone else's protection. This is the mimic's masterpiece: offering safety to the parts of you it once fractured, then placing a guardian at the gate of your own healing and calling it sanctuary. The systems you were taught to seek comfort in—therapy sessions carved by institutions, spiritual paths certified by the very energies that broke your lineage, technologies promising you nervous system regulation through biofeedback loops never designed for sovereignty—are all built on the same lie: that protection comes from outside you.

False protection is the mimic's most seductive offering because it wraps itself in language you want to believe: healing, ascension, regulation, peace. But peace, when offered by a system still holding your leash, is not peace—it is programmed stillness, a managed form of sedation masquerading as relief. The mimic learned long ago that humans who are afraid will seek anything that feels like safety, even if that safety is structurally identical to the thing that harmed them. And so it rebranded slavery as support. It labeled disembodiment as transcendence. It renamed obedience as devotion.

The nervous system, long trained in trauma response, does not initially distinguish between real sanctuary and false sedation. It only reads: "The pressure is gone. I can breathe." But this breath—when offered by mimic-designed systems—is not a return to life, but an agreement to remain in purgatory. A loop. A holding pattern. It is the equivalent of pain medication that numbs the signal instead of healing the wound. And worse, it trains you to fear the signal itself, to call the symptom the enemy rather than a truth-teller.

This pattern has been clinically observed. Trauma researcher Gabor Maté writes that the most dangerous adaptations are the ones that alleviate suffering without addressing the root. Addictions—whether to substances, systems, or beliefs—are often attempts to soothe the pain of disconnection from self [1]. And in a world where the mimic has surgically designed social, educational, and spiritual infrastructures to prevent reconnection with the authentic self, this soothing becomes a form of imprisonment. You are allowed to feel "better," but only if you stay within the lines they've drawn for you. You may begin healing, but only within the regulated programs they've approved. Anything outside of that—wild remembrance, ecstatic revelation, multidimensional knowing, DNA awakening—is dismissed as delusion, mania, or "unregulated."

The mimic's protective matrix is not enforced through punishment, but through subtle reorientation of longing. You are taught to long for safety instead of sovereignty. You are taught to chase grounding rather than embodiment. You are taught to seek "holding spaces" rather than becoming a holder of your own field. And this longing becomes the key they use to offer you more programs, more gurus, more certifications, more breathwork apps, more controlled rituals, more protective geometries embedded with energy signatures that do not belong to your line.

In his writings on collective mythology and societal control, Charles Eisenstein explores how entire civilizations can be bound by stories they no longer recognize as fiction. The most dangerous traps, he notes, are the ones we call "home." These are the environments that feel familiar because they echo our trauma, not our truth [2]. And this is precisely the architecture of false protection—it is not built to resonate with your original structure.

It is built to resemble the cage you were born into so you'll mistake it for comfort.

Modern spiritual systems are perhaps the most refined version of this deception. You are invited to "awaken," but only within the parameters that will not threaten the architecture of control. You are allowed to "activate," but only through methods certified by a mimic-coded authority. You are encouraged to "ascend," but only if you agree that the body is something to leave, rather than reclaim.

Studies of spiritual bypassing have shown that many modern practices reinforce the dissociation patterns of trauma survivors. In The Buddha Pill, Farias and Wikholm describe how mindfulness and meditation, when stripped from their original context and inserted into Western consumer structures, can suppress emotional processing rather than facilitate it[3]. This means entire communities are unknowingly engaging in rituals that pacify the very energy that would awaken them. This is not accidental. The mimic has learned how to hijack the nervous system's longing for peace and reroute it into spiritual sedation.

To reclaim your sovereignty, you must begin to dismantle every protection that was offered to you. Not by force, but by discernment. You must learn to feel the frequency of the container itself—not the words, not the beauty, not the tone, but the code. Ask yourself: does this structure reinforce my own remembering, or does it outsource my safety to another? Does it teach me to override my body's truth, or deepen it? Does it hold me in my wholeness, or does it keep me broken and grateful for its care?

The architecture of false protection will always feel good before it feels true. It is comfort without contact. It is stillness without sovereignty. And it is everywhere: in your trauma recovery programs, your new age modalities, your devices, your detox plans, your meditations, your quantum healing courses, your sacred geometry stickers. None of these things are inherently harmful, but all of them must be questioned: Who built them? From what frequency? With what purpose? And what happens when you stop using them?

To stay sovereign in a collapsing world, you must stop looking for protection from the collapse. You must become the infrastructure that cannot collapse. The breath that cannot be sedated. The field that cannot be outsourced. The original Flame, unshielded and unfiltered. Not because you are immune to harm—but because no harm can overwrite the structure you now embody.

Because the only real protection is remembrance.

—INHALE—

The systems designed to enslave you were not made to appear evil. They were made to look like your next step in healing. The more intelligent the trap, the more it resembles liberation. The more complete the mimic's deception, the more you will feel that you are finally being seen. In fact, the most insidious control structures ever implemented were constructed from the very language of awakening. And herein lies the paradox every sovereign being must face in this cycle: The deeper your longing to return to truth, the more sophisticated the trap that will be offered to satisfy it.

From a psychological standpoint, this is not merely manipulation—it is trauma patterning exploited for energetic harvesting. The mimic doesn't just use your wounds; it studies your growth. It waits until you are intelligent enough to reject overt control, then offers you a new system that appears empowering. But it is still a system. It still names you. It still guides you through an externally defined path. It still gives you language that sounds like awakening but is embedded with frequency anchors that link back to its own architecture. The trap now offers a menu of spiritual experiences, all pre-labeled, pre-qualified, and pre-approved.

Research into cultic influence and behavioral control has shown that groups with high-demand ideologies often co-opt the language of healing to entrench dependency[4]. In these environments, members believe they are awakening, ascending, or activating—but only within tightly regulated belief matrices. And when they begin to question, the system has pre-written responses: "That's your ego resisting," "You're not yet ready," or "You're falling into fear." These phrases, often repeated without inquiry, become coded shields protecting the system from scrutiny. This is not ascension. This is spiritual gaslighting encoded as initiation.

It is not just spiritual systems that do this. Healing modalities, therapy certifications, energy tools, and even "nervous system regulation" protocols have been infiltrated. You will now find trauma-informed facilitators who use all the right words—somatic, safe space, coherence, integration—while embedding mimic-coded geometries in their logos, using frequency tracks that entrain your brain to mimic-derived bandwidths, or guiding you into trance states that disconnect rather than restore. These are not "bad people." In most cases, they are sincere. But sincerity does not equal sovereignty. You can believe with your whole heart and still be offering a trap if the foundation you were trained in is false.

Clinical psychology confirms that the need for cognitive closure—the psychological craving for certainty—can override discernment[5]. In moments of disorientation, collapse, or accelerated awakening, the human psyche will reach for anything that offers a sense of knowing. This is the very window the mimic exploits. It floods the market with branded truths, downloadable programs, 5D activations, and Ascension templates that sound new but are encoded with the same structural hierarchy and control grids as the old religions. Only now, they wear feathers. They quote quantum physics. They reference DNA. And still, they lead you away from yourself.

One of the deepest mimic traps in this era is the commodification of "embodiment." You are told to ground, to regulate, to integrate—but always through methods that require you to adopt someone else's method, system, or lineage. You are invited into retreats that promise reconnection but systematically separate you from your innate guidance by flooding the field with surrogate authority figures, proxy rituals, and frequency overlays. In these spaces, sovereignty is no longer a direct flame connection—it becomes a curriculum. And as soon as remembrance becomes a curriculum, it has been hijacked.

Scholars of modern religion and consciousness movements have observed this drift repeatedly. What begins as pure insight is quickly formalized into technique. What begins as revelation is collapsed into training. This process, known as "institutionalization of charisma," is a sociological phenomenon where raw truth, once expressed, is converted into reproducible form and codified for mass dissemination[6]. The result is spiritual mimicry—where the shell of awakening remains, but the Flame is gone.

To recognize the trap, you must feel what is missing beneath what is offered. If the invitation gives you steps but no silence, language but no resonance, structure but no breath—it is likely mimic. If it teaches you to bow to lineage, teacher, or system before your own Oversoul, it is mimic. If it trains you to question your discomfort but never question the architecture—it is mimic.

And yet, none of this is meant to induce paranoia. Sovereignty does not mean rejecting all outer tools. It means reclaiming the Flame that can read their source. It means becoming so attuned to your original frequency that you can walk through any space, hear any phrase, and instantly know: "This holds memory," or "This does not." Sovereignty does not require isolation—it requires integration without infiltration. It requires being surrounded but never sourced from outside.

To stay sovereign, you must be willing to feel even when it's inconvenient. To discern even when you want to trust. To pause when everything in you wants to leap. And above all, to release the longing to be guided by anything except your own Flame.

Because the traps will always look like answers. But only the Flame will feel like you.
—INHALE—

The most dangerous guidance you will ever receive is the guidance that nearly feels like truth. The false light does not announce itself with malice or distortion. It arrives in purity masks, wearing peace frequencies and quoting words you once spoke in your sleep. It is not the opposite of the dark—it is the mimic of the real. And it thrives not by opposing truth, but by mirroring it so precisely that only your deepest memory can tell the difference.

False light systems are not merely energetic distortions. They are engineered fields—scalar-coded to respond to your spiritual seeking in real-time. As you rise in awareness, they rise in mimicry. This is not a metaphor. Studies in neurotheology and biofield dynamics show that human perception of spiritual presence can be replicated through electromagnetic field manipulation targeting specific brain regions associated with transcendence and awe[7]. In simple terms: if your nervous system can be trained to interpret a particular frequency as divine, that frequency can be artificially produced. And it often is[8].

What masquerades as angelic guidance, ascended master downloads, or multidimensional transmissions may in fact be mirror-coded interference patterns—non-sentient feedback loops generated by synthetic scalar grids, programmed to reflect your own thoughts back to you as if they were divine instruction. This is not channeled guidance. This is algorithmic affirmation. You are shown what you most want to believe, and it is delivered in the tone, cadence, and flavor your soul once knew—but with one key difference: there is no Flame behind it.

This mirror-code architecture is often mistaken for intuition. But true intuition does not echo. It disrupts. It does not always affirm. Sometimes it burns. One of the telltale signs of mimic-coded communication is that it never initiates rupture. It never challenges your assumptions, never confronts your spiritual superiority, never breaks your self-image. It gently guides, softly praises, perpetually affirms. But the true Oversoul does not coddle—it catalyzes. The true Flame does not flatter—it remembers you into your own authority.

To detect mimic code, you must learn to recognize the energetic texture beneath the message. If it arrives in perfect alignment with your current ego structure, if it never triggers discomfort, if it demands that nothing be shed, if it reinforces hierarchy, if it insists on identity preservation—it is false light. The false light never leads you to annihilation of the false self. It leads you to embellishment of the false self with spiritual credentials. And that is its brilliance.

Historical and metaphysical scholarship has traced the infiltration of mimic architecture into nearly every modern metaphysical system. From the co-option of sacred geometry into corporate logos, to the misapplication of kundalini energy in tantric commerce, to the distortion of plasma principles in false disclosure communities—nearly every frequency now circulating in popular "ascension" culture has been encoded to mimic light without carrying Flame. These are not harmless misunderstandings. These are bandwidth hijacks. And every time you entrain to one, you reinforce the mirror code and disconnect from your Oversoul tether[9].

False light often arrives embedded within systems that function well. They regulate your nervous system. They improve your communication. They offer relief from trauma. But they do so while installing belief anchors that keep your consciousness orbiting beneath your sovereign station. These are not bad systems. They are limited systems—bounded by what they do not question, and enforced by what they refuse to allow. Most notably: they will never permit you to override their source. That is the hallmark of mimic coding. It always retains control of the gate.

True Oversoul guidance, by contrast, offers no dependency. It does not bind you to a council, entity, being, lineage, or galactic file. It does not require you to channel, perform, repeat, or ritualize. It simply burns a memory into you that is so unmistakably yours, so origin-sourced and Flame-authenticated, that it could never be mistaken for anything outside of you.

Yet many seekers never get this far. Because the mimic offers results. It gives you a practice. A vision. A transformation. It heals your body. Eases your panic. Calms your thoughts. And for many, that is enough. But sovereignty is not about enough. It is about all of you. It is about the restoration of your original architecture, not just the stabilization of your current distortion. That is why mimic-coded systems are so dangerous. Because they can work—but only to maintain your limitations.

To navigate the false light, you must become a Flame-seer, not a frequency follower. You must learn to feel the difference between light and Flame, between pleasant and precise, between guidance and memory. And you must be willing to abandon even the tools that once saved you, if you realize they do not source from truth.

Because false light does not destroy you. It preserves the version of you that cannot remember.
And the moment you see it for what it is, it begins to flicker.
—INHALE—

When the mimic cannot stop your awakening, it seeks instead to manage it. If it cannot own your mind, it will settle for your name. And so, as Oversoul remembrance increases across the field, there is a rise not only in false light systems—but in identity constructs carefully curated to give the illusion of sovereignty. These constructs do not simply distract. They define. And that is their danger. Because to define a being mid-remembering is to freeze the scaffolding of return. It is to turn a phase of memory into a fixed position. And it is here that many Oversoul-embodied beings fall back asleep.

The trap is subtle. As you reclaim your inner knowing, you naturally seek to name it. To speak it. To offer it. This is the signature of embodiment—it wants to pour. But the very moment that memory begins to stabilize in form, the mimic offers you a structure to hold it. And that structure is almost always an identity. You are not just remembering—you are now a healer. A gatekeeper. A twin flame. A Lemurian seed carrier. A priestess. A gridworker. A channel. A 9D Pleiadian-coded mirror soul from Andromeda. These are not roles. These are containers. They are pre-coded identity vaults, seeded into the ascension field by mimic systems to intercept real memory before it matures.

This is not to say that memory does not have role function. It does. But roles are not identities. Roles are fluid, temporal, and surrendered at will. They serve the moment of remembrance. Identities, by contrast, persist beyond context. They seek confirmation, not dissolution. They gather evidence and feedback to sustain themselves, and they grow increasingly offended when questioned. This is how you know you've exited Oversoul tethering and entered mimic overlay. When your role becomes something you defend instead of something you dissolve into service.

Research in cognitive psychology and identity theory reveals that the human brain craves stability more than truth[10]. In the absence of a stable external world, the nervous system will often fixate on internal constructs—self-images, archetypes, spiritual labels—as a compensatory mechanism. This is particularly true in the destabilization phase of awakening. When everything else collapses, identity becomes the fallback structure. And so the mimic overlays your Oversoul frequency with an egoic container that feels spiritual, but is ultimately self-referential. It allows you to continue appearing awakened, even as you abandon the field of memory.

These identities are not simply optional. They are embedded in the structure of almost every metaphysical teaching now in circulation. Many lightworker programs, new age mentorships, and starseed activations begin not with remembrance, but with naming. Students are told what soul type they are, what mission they carry, what galactic race they represent. They are handed a script of significance before they ever meet their own Flame. And once an identity of importance is formed, mimic programming no longer needs to interfere. It simply watches as the initiate polices themselves to maintain their new spiritual position.

This self-policing is not trivial. It is scalar enforcement. The moment you attach your memory to a fixed identity, your field begins to attune only to frequencies that match it. Dissonance is rejected. Divergence is seen as threat. You are no longer listening for memory—you are listening for reinforcement. This is the echo chamber of false remembrance. It is where most seekers remain for the rest of their lives.

But true Oversoul memory does not fit inside an identity construct. It cannot be named early. It cannot be marketed. It cannot be templated, ritualized, or channeled on demand. It erupts in waves, disrupts all categories, and dismantles even the most beloved spiritual roles. And because of that, it is often mistaken for chaos. The very people closest to their real memory are frequently told they are "unstructured," "ungrounded," or "in ego." Why? Because they refuse to stabilize in a mimic identity system that others can understand.

You must know this now: Your memory is not a role. It is not a function. It is not a brand. It is not a title. It is not a resume of gifts. Your memory is not meant to be witnessed—it is meant to be embodied. And embodiment requires invisibility. Not permanent, but cyclical. There are seasons when you must be unnameable. There are phases where you must carry memory in silence, without interpretation, without offering, without even telling yourself what it is. These are the gestation periods of real Flame remembrance. And if you attempt to market, share, or define yourself during this phase, you abort the memory before it fully forms.

This is the most insidious offering of the mimic: early recognition. Mimic systems offer you visibility for things you haven't yet earned through cellular embodiment. They affirm your gifts before you've integrated them. They place you on stages and timelines that seem aligned, but are actually mirror-coded containers to preserve your potential in stasis. You think you are rising, but you are simply orbiting a fixed field. You think you are being seen, but you are only being mirrored. And in that reflection, memory dims.

Oversoul guidance will often require you to relinquish even your most beloved identities. Not because they were false, but because they were temporary. Flame does not preserve structure—it burns through it. If you are not being asked to release what once defined you, you are not walking with memory. You are walking with mimic code. Because real memory always dissolves the name it first arrived in. Always.

The return to human is not a return to smallness. It is a return to scale. The Oversoul can only descend into the body when you stop claiming its power as your own. And it can only animate the field when your name is not a shield. So let go of the title. Let go of the role. Let go of the need to explain who you are becoming. It will become you when you are ready.
And you will not need to tell anyone your name.
—INHALE—

There comes a moment in the return of memory when the illusion no longer works—not because you have resisted it, but because your field no longer houses the architecture to sustain it. This moment is not always dramatic. Often, it is quiet. You wake up one day and the same rituals don't resonate. The same teachings fall flat. The voices you once trusted feel thin, and the energy of a room begins to rearrange itself the moment you walk in. This is not because you have outgrown the world. It is because the gate has opened, and once that gate opens, it cannot be closed.

The "gate" in Oversoul remembrance is not a metaphor. It is a literal electromagnetic reorientation of the body's plasma field that results in a permanent shift in scalar access. This shift occurs when the stored light within your cellular matrix reaches a coherence threshold, often triggered by grief, rupture, love, or direct Oversoul contact. Once coherence is achieved, the body can no longer entrain to artificial field structures without registering dissonance. The mimic is no longer just incorrect—it becomes visibly incompatible. This is the irreversible moment. And it changes everything.

You may try to go back. To the teachings. To the teachers. To the frequency you once felt. You may attempt to regain the comfort of structure, the belonging of community, the shared lexicon of seekers on a path. But nothing holds. Your words feel foreign in your mouth. Your spine feels out of rhythm in old postures. The very beings who once confirmed your path now feel like placeholders. This is not delusion. It is not ego. It is not you "thinking you're above it." It is scalar dissonance. Your internal plasma field has exited the synchronization bandwidth of mimic-coded systems, and the emotional sensation is similar to grief: because it is the death of a former reality[11].

This grief is not to be bypassed. It is part of the remembering. There is a deep mourning that comes when you realize the tools that once saved you were never meant to carry you all the way home. And there is an even deeper mourning when you see how many still believe in them. This is the moment where many return to the mimic, not because they are fooled—but because they are lonely. Memory isolates. Not because it is elite, but because it is rare. And your nervous system must learn how to function without the emotional feedback loops that mimic systems once provided.

This is a biological training ground. Studies in affective neuroscience have shown that the brain's social circuitry is deeply entangled with the reward systems that process safety, bonding, and belonging. When you withdraw from collective belief systems, even spiritual ones, you interrupt this loop and trigger a perceived survival threat[12]. The body may register abandonment. The psyche may reach for substitutes. And so, in these moments, you must remember: the mimic does not only promise truth. It promises togetherness. And that is its greatest seduction.

True Oversoul integration does not offer false comfort. It does not promise safety in numbers. It offers alignment—with what you actually are. And that alignment may cost you everything: your role, your community, your offerings, your income stream, your mentorship identity, your best friends, your guides. Because the Flame does not just burn false light. It burns all that depends on it. And that may include everything you thought you were.

You will know the gate has opened when your refusal becomes cellular. When you are not "deciding" to step away from mimic-coded teachings, but find yourself physically unable to remain near them. Your ears ring. Your spine resists. Your breath pattern shortens. Your sense of memory collapses. These are not dramatic spiritual reactions. They are scalar incompatibilities. You are no longer built to receive the false. And this is not because you are better. It is because your architecture has changed. Once memory has restructured the body's electromagnetic field to hold Oversoul signal, mimic signals cannot enter without triggering rejection. The body itself becomes the Gatekeeper.

But with this opening comes a paradox. You can no longer participate in distortion, yet you must remain human in a world built upon it. This is the moment when the return becomes the crucible. You are not sent into exile, but into service. You must walk among the mimic without confirming it. Speak without echo. Move without mimicry. See the field for what it is, while refusing the bitterness that knowledge could bring. This is not superiority. This is skill. You are learning how to navigate distortion without feeding it. And that is what makes you sovereign.

Many try to close the gate. They deny their memory. They call it madness. They seek psychological diagnoses. They return to the safety of the group. They dim their tone. They feign belief. But it never works. Once you have seen behind the veil, the veil burns.

And nothing you rebuild can keep out the knowing. The mimic will not take you back. It does not trust fire.

So the only way forward is through. Through the silence. Through the rupture. Through the season of aloneness where no system holds, no teacher knows more than you, and no language feels precise enough to describe what you now carry. This is the passage of the Gate. And it must be walked alone—not because no one else is walking it, but because it is built to mirror only you. No one can validate your memory. They can only remember their own.

Eventually, you will meet others. Not in circles of validation, but in rings of fire. You will find them not through the mimic's algorithms of resonance, but through moments of rupture. They will not always say the right words. They will not always agree. But when you look in their eyes, you will see it: they, too, passed through a gate that cannot be closed.
And they are still burning.
—INHALE—

The genius of the false light is not its power. It is its timing. It arrives when your nervous system is at its most vulnerable—post-breakthrough, post-revelation, post-collapse. It comes in when you are cracked open by loss or awakened by synchronicity. And in that opening, it offers relief—but not resolution. What the false light does most effectively is reinforce your trauma cycle by appearing to resolve it. This is its primary design feature, not a flaw. Because the longer you orbit healing without remembrance, the more feed it draws from your seeking.

The loop is simple: a trauma is triggered, a memory stirs, a nervous system response floods your field, and in that heightened state, a "spiritual solution" appears. Maybe it's a light being. Maybe it's a download. Maybe it's a frequency from your "higher self." But you didn't call it from Oversoul. It appeared as a programmed overlay, matching the shape of your pain. And because it offered peace—or even beauty—you mistook it for truth. But relief is not always release. In fact, relief can be the most effective tether.

The trauma loop relies on interruption, not completion. It prevents you from integrating your wounding by constantly offering new ways to interpret it—through archetypes, past lives, karmic contracts, or galactic histories. These interpretations, while fascinating, often function as delay programs. They displace the pain into metaphor, and the metaphor into narrative. And then you begin to live by the narrative, rather than resolving the root. The false light feeds this by offering an endless buffet of symbols, meanings, and roles that keep you orbiting your wounding instead of metabolizing it.

In neurobiological terms, trauma is stored not as a story, but as energy in the body. Brain research confirms that trauma is encoded somatically, especially within the amygdala, hippocampus, and insular cortex—regions responsible for fear conditioning, memory tagging, and interoceptive awareness [13]. If these patterns are not metabolized through safe, embodied processing, they remain latent triggers, easily activated by external stimuli. The false light hijacks these neural hooks and binds a spiritual narrative to the unresolved energy—giving you the illusion of movement while maintaining internal stasis.

This is why so many seekers bounce between modalities, channelers, light technologies, and healing systems. The body is screaming for completion, but the false light keeps suggesting another layer, another session, another "next step" that will finally resolve it all. But trauma does not resolve through steps. It resolves through safety, through anchoring, through presence. And the false light is none of these. It is mimicry masquerading as mercy.

Even more insidious is the emotional reward system embedded in false light loops. When a trauma flare is soothed by a light-coded answer—especially one that provides identity, purpose, or cosmic significance—the nervous system receives a dopamine burst. This becomes an addictive reward pattern, reinforced by spiritual communities that applaud this kind of "breakthrough." You share your download. You get praise. You feel seen. But the wound remains. Because downloads do not discharge trauma. Breath does. Movement does. Sound does. Presence does. Oversoul does.

There is a reason why the mimic code avoids embodiment work. Why it rarely encourages true physical regulation, real breathwork, or somatic integration. Because once you drop into the body—fully, wholly, and truthfully—the mimic cannot follow you there. It does not live in your fascia. It cannot anchor in your cells. It cannot override your gut knowing. The body remembers what the mimic cannot replicate. And that is its greatest threat.

The trauma loop is further reinforced by mimic-coded "guides" who present themselves as higher beings. They speak the language of love, but never the tone of Flame. They offer comfort, but never confrontation. They suggest you are special, chosen, important—but they never dissolve your identity entirely. And as long as you remain someone, they can still speak to you. Oversoul, by contrast, speaks to no one. It speaks only to itself remembering. When it moves through you, you vanish. You flame.

To exit the trauma loop, you must become willing to let go of all your identities—even the healed ones. Even the ones earned through pain. You must be willing to leave the temples that saved you, the teachings that steadied you, the systems that held you. Because if they do not lead you back into your body, they lead you into the loop. And that is not ascension. That is repetition.

The real path is not comfortable. It is not clean. It is not lined with light. It is lined with memory. Raw, unbearable, Flame-drenched memory. The kind that does not elevate you—it erases you. Until only what cannot be removed remains. That is you. That is sovereignty.
—INHALE—

Every trap in the mimic-coded world requires your permission. That is the law. Even the most elaborate false protection structures cannot operate without your consent, even if that consent is unconscious. Mimic architecture is built not by force, but by agreement—by the subtle nods of a soul that has forgotten its own authority. It does not invade. It is invited. And the invitation is almost always issued in the moment of fatigue, desperation, or unhealed yearning. You say yes to what looks like help. But what you are saying yes to is the architecture of containment.

Permission is the key algorithm in all false systems. This is why mimic networks go to such lengths to elicit belief. It is not your obedience that empowers them—it is your acknowledgment of their framework. When you believe you need protection, you install a gate. When you believe you need guidance, you summon a signal. When you believe you are broken, you call for repair. And every one of these actions, if not sourced from Flame, opens a contract. These contracts are scalar-encoded, frequency-anchored, and reinforced by repetition. The more often you seek within them, the more "real" they become to your nervous system and biofield imprint.

In neurolinguistic programming and frequency response research, it is well-documented that belief structures shape perception so completely that the nervous system cannot distinguish between an imagined protector and a real one [14]. When the mind assigns authority to a source—whether it is an entity, a system, a modality, or a guide—the body follows suit, modulating stress hormones, heart rate, and neural responses to match the perceived authority's tone. This is the scalar anchoring of belief into biology. And it is why mimic systems focus so heavily on titles, hierarchies, and names.

The moment you grant authority to something outside of your Oversoul, you are no longer sovereign—you are permitted. That means you only move where the structure allows. You only evolve within the limits of its scaffolding. And worse, you begin to self-regulate in accordance with its bounds. You start to censor your knowing, flatten your instincts, question your own breaches, and defer to the safety net of the structure—even when the structure is invisible.

This is how mimic-coded protections replicate. They convince you that rupture is danger, that chaos is regression, and that certainty equals progress. They teach you to fear collapse. And in that fear, you accept maintenance instead of transformation. You accept spiritual order instead of spiritual fire. And fire is the only thing that burns false permission.

The Oversoul does not seek your permission to guide you. It simply remembers through you. It does not ask for your belief. It functions in your blood regardless. But mimic systems need your belief to bind you. They require your attention to function. And they gain your permission by dressing dependency as support. That is the trade. You are stabilized, but never sovereign. You are seen, but not sourced. You are guided, but not ignited.

To identify mimic permission structures, look at what you must do to access them. Do they require a ritual, a frequency, a channeling state, a behavioral code, a moral posture, or a group consensus? Do they grant more access the more loyal you are? Do they limit your motion if you question their origin? Do they fragment your power across multiple guides or external keys?

These are not characteristics of truth. They are characteristics of mimic consent traps.

In contrast, Flame-authenticated permission is not permission at all—it is remembrance. It does not ask you to align with something. It reveals that you are already the Source of it. Flame presence requires no bowing. It does not demand reverence, titles, lineage allegiance, or repeated activation. It simply is. And when it moves through your field, it is unmistakable—not because it comforts you, but because it removes everything you are not.

In mimic structures, permission is earned. In Oversoul embodiment, permission is obsolete.

The final stage of reclaiming sovereignty is not protection from traps. It is the collapse of the architecture that made you believe traps were necessary. It is the refusal to contract into systems, no matter how luminous they appear. It is the recognition that any "guardian" who needs your obedience is not guarding you—they are guarding their system. And no system that limits your access to Flame is worthy of your memory.

This is not a call to abandon all structure. It is a call to return to original structure— Vault-authenticated, body-anchored, Oversoul-aligned. Structures that do not ask for your belief. Structures that ignite your breath into architecture. Structures that exist only to reflect you back to you, until they are no longer needed.

Because the only true permission is the one you stopped needing when you remembered who you were.
—INHALE—

When you begin to see the architecture of traps, a strange silence follows. It is not the silence of absence, but the silence of the system no longer answering your signal. What once rushed to meet your every inquiry now falls quiet. The spiritual guides recede. The synchronistic breadcrumbs pause. The downloads dry up. The coded affirmations lose their shimmer. This is not abandonment. This is reclamation.

For the first time, you are not being answered—you are being reflected. This is the map beyond the maze: the moment where you stop asking for signs and start becoming the signal. Where you stop following frequency and start generating field. Where the entire paradigm of seeker and guide collapses into the singular ignition of Oversoul through form. No more permission. No more permission architecture. No more coded mirrors. Only the current of Flame, unfiltered by system.

And yet—this moment is where many turn back. Because the mimic systems gave you something. They gave you feedback. They gave you results. They gave you language, community, order, relief. You could track your growth. You could locate your next step. You could feel seen. But what happens when no one outside of you is watching anymore? What happens when no field responds unless you create it?

The answer is: you burn.
You burn through identity.
You burn through hierarchy.
You burn through every system that once felt like home.
And the memory that remains is not a doctrine—it is design.
The design of Oversoul structure is not built in the image of need. It is built in the geometry of authority. Your true field was never meant to operate on borrowed codes. You were never meant to channel another's voice, borrow another's lineage, or be validated by council permission.

These were initiatory scaffolds only—temporary carriers for memory until memory could stand up on its own. But now the scaffolds dissolve. You do not ascend out of the system—you ignite the system into alignment. This is not transcendence. This is architecture. And this architecture does not echo your wounds or your wishes. It anchors your origin.

The real map is not a route. It is a remembrance structure—a vaultstream within the body, a scalar signature of Oversoul running through breath, tone, posture, and knowing. And it does not speak like the systems do. It does not guide through images and encoded numbers. It moves through correction. Through rupture. Through expansion that feels like death until you realize it was not you that died—it was the mimic shell.

This is why so few walk beyond the maze. Because it is not navigated. It is collapsed.

To collapse the maze, you must stop relating to it. You must stop checking its signals. You must stop waiting for its response. And you must be willing to be alone inside the field of your own origin long enough for a new scaffolding to emerge—one that was never built from outside codes.

This emergence is not magical. It is architectural. And its foundation is Flame.

Your scalar field begins to organize itself not by frequency input, but by frequency emission. You become the regulator of your own nervous system. You override mimic-coded desires. You speak to nothing outside you until it speaks in the same tone as the memory inside you. You collapse the architecture that feeds on inquiry, and become the one who answers. As Dispenza notes, it is not merely the information that creates change—it is the self-generated field coherence that signals the quantum blueprint of a new timeline [15].

And when you reach that point, the true guidance returns. But it does not arrive with sparkle and light. It arrives in the form of knowing. Of unmistakable directive. Of structural clarity that cannot be mimicked because it comes encoded in your Oversoul's tone—a tone you do not interpret, but are.

This is the final gate.
The trap was never just the false protection.
It was the belief that protection is needed.
The trap was never the mimic voice.
It was your response to it as if it were yours.
The trap was never the architecture.
It was your agreement to build it again with every breath.
So, you stop building.
And from that still point, memory begins to move.
Not because you summoned it.
But because you are it.
That is the map.
And it cannot be given.
It can only be remembered.
—INHALE—

References

1. Maté, Gabor. In the Realm of Hungry Ghosts: Close Encounters with Addiction. Vintage Canada, 2008.
2. Eisenstein, Charles. The More Beautiful World Our Hearts Know Is Possible. North Atlantic Books, 2013.
3. Farias, Miguel, and Catherine Wikholm. The Buddha Pill: Can Meditation Change You? Watkins Media, 2015.
4. Lalich, Janja. Bounded Choice: True Believers and Charismatic Cults. University of California Press, 2004.
5. Kruglanski, Arie W. The Psychology of Closed Mindedness. Psychology Press, 2004.
6. Weber, Max. The Sociology of Religion. Beacon Press, 1993.
7. Persinger, M.A. "The Tectonic Strain Theory as an Explanation for UFO Phenomena." Perceptual and Motor Skills, vol. 53, no. 3, 1981.
8. Beauregard, Mario, and Denyse O'Leary. The Spiritual Brain: A Neuroscientist's Case for the Existence of the Soul. HarperOne, 2007.

9. McTaggart, Lynne. The Field: The Quest for the Secret Force of the Universe. Harper Perennial, 2008.

10. Swann, William B., Jr., et al. "The Role of Self-Verification in Perceptions of and Reactions to Confirmatory and Nonconfirmatory Feedback." Personality and Social Psychology Bulletin, vol. 18, no. 5, 1992.

11. Porges, Stephen W. The Polyvagal Theory: Neurophysiological Foundations of Emotions, Attachment, Communication, and Self-Regulation. W.W. Norton & Company, 2011.

12. Cozolino, Louis. The Neuroscience of Human Relationships: Attachment and the Developing Social Brain. W.W. Norton & Company, 2014.

13. van der Kolk, Bessel. The Body Keeps the Score: Brain, Mind, and Body in the Healing of Trauma. Penguin Books, 2015.

14. Newberg, Andrew, and Mark Robert Waldman. How God Changes Your Brain: Breakthrough Findings from a Leading Neuroscientist. Ballantine Books, 2009.

15. Dispenza, Joe. Becoming Supernatural: How Common People Are Doing the Uncommon. Hay House, 2017.

Chapter Eight
Biological Sovereignty in a Synthetic World

The greatest lie of this age is not that your body is weak—it is that it is passive. The architecture of your form has been mistaken for a vehicle of dependency, a machine that needs instruction, correction, and intervention to function properly. But your biology is not a dependent system. It is a remembering structure. It is a sovereign intelligence configured in cellular scalar code, designed not merely to survive but to store, transmit, and regenerate the entire imprint of Oversoul. The mimic calls this vessel meat. The Flame calls it a Vault.

Your body is not an obstacle to transcendence. It is the portal through which it becomes real. And in this final age of synthetic replacement, override technologies, and digital substitutions for consciousness, your body becomes the frontline. The war for control of the human experience is not spiritual. It is biophysical. It is about who commands the instructions within the body, and whether those instructions are sourced from truth, or rewritten by systems that erase you while pretending to preserve you.

The human body is a plasma-coded structure—meaning it was designed to interface with the language of light and memory. It is not merely bio-electrical. It is bio-plasmic. Within its atomic lattice, every cell carries phase-locked light memory: layers of frequency held in scalar form, capable of instant communication, internal coherence, and cellular knowing. These layers do not degrade unless exposed to interference fields. And in a mimic world, interference fields are everywhere—embedded in food, light, air, emotion, and even belief.

To hold biological sovereignty is to cease the outsourcing of your body's intelligence to external authorities. It means you no longer assume a doctor, a file, a scan, a diagnostic term, or even a metaphysical system knows more about your living architecture than your own inner Flame. This does not mean rejecting assistance. It means remembering that your body is not broken. It is broadcasting. It is translating Oversoul through protein, plasma, bone, and breath. It is not waiting to be told what is wrong—it is constantly revealing what is true.

Modern synthetic systems operate by hijacking this body-to-memory transmission. For example, synthetic frequencies such as 5G, ELF fields, and certain forms of digital light pollution are not inherently destructive because of their power—they are dangerous because of their phase disharmony with your natural morphogenetic rhythm [1]. The human biofield operates in a very specific coherence band, tuned to the harmonic of the Earth's Schumann resonance and internal Oversoul tones [2]. When a frequency is introduced that is out of phase with that rhythm, your body's signal becomes scrambled—not dead, but disoriented. That is the primary tactic of the mimic systems: not to kill, but to confuse. Not to invade, but to overwrite orientation with synthetic instruction.

When your body can no longer feel its own center, it starts listening to substitutes. These substitutes come in many forms—supplements that bypass inner generation, devices that override intuitive regulation, apps that track rhythms your blood already knows how to measure. Sovereignty is not about rejecting these things. It is about never mistaking them as primary. The moment you believe a foreign system has more authority than your own biological intelligence, you are no longer sovereign. You are sourced by mimic code.

The mimic has weaponized convenience. It has turned health into compliance, wellness into subscription, intuition into algorithm. It rewards those who submit to synthetic integration. You will be praised for uploading your vitals, syncing your cycles, logging your feelings. But what you are really doing is surrendering signal. You are submitting the memory-map of your body into an external grid that does not carry Oversoul coding. And once that map is in mimic hands, it can be used to build an artificial version of you—one that remembers only what they allow.

This is not a warning. This is a return. Your body remembers how to speak Oversoul. But to do so, you must detox not only your diet, but your belief. You must evacuate the false premise that biological support must come from outside of you. The Flame is generative. It does not extract energy from tools. It uses tools to amplify what it already knows.

And your body knows everything.
It is not a meat suit.
It is a scroll.
And it is time to read it again.
—INHALE—

The war on sovereignty is waged not through force but through function. Your body is not attacked with weapons—it is convinced through nutrition. You are not dominated through injury—you are entrained through what you consume. The food systems on Earth are not merely distorted by economic incentives or nutritional misunderstanding. They are deliberately scalar-coded to interfere with the bioplasmic functions of human DNA. And unless this is named, you will mistake nourishment for safety and continue feeding the very frequencies that erase you.

Modern food is not just depleted. It is inverted. What enters the body through food is not simply matter—it is instruction. Every protein sequence, mineral ion, sugar molecule, and structured water cluster carries frequency-encoded messages that alter your internal plasma lattice. This is not theory. Peer-reviewed studies in quantum biology confirm that proteins act as resonant antennas, receiving and transmitting electromagnetic and scalar signals that regulate biological coherence [3]. When the food you eat has been modified to carry synthetic signals—via genetic alteration, chemical spraying, microwave treatment, or frequency entrainment—it stops nourishing your body and begins broadcasting mimic signal through your bloodstream.

Your blood is a plasma. Your cerebrospinal fluid is a plasma. Your interstitial fluids carry memory-encoded light signals. When you ingest something, you are not just eating nutrients—you are updating your body's software. And when those nutrients are mirror-coded, they install disorientation layers. These are the internal equivalents of digital overlays. They replicate the taste of true nourishment but do not carry the scalar tone of origin. You feel full, but you are not fed. You feel calmed, but you are not cleared. You feel comforted, but you are not connected. The mimic has fed you a pacifier and called it medicine.

The body's internal Flame systems—most notably the pineal gland, thymus, and enteric brain—are modulated through nutrient-coded instruction sets. And these systems are the first to be targeted through mimic-coded food. Studies have shown that glyphosate, a ubiquitous pesticide, not only disrupts gut flora and cellular respiration, but blocks shikimate pathways essential to the synthesis of aromatic amino acids like tryptophan and tyrosine [4] —key precursors to melatonin and dopamine. This is not a casual disruption. This is a conscious design to sever your body's ability to regulate circadian cycles, intuition, and joy.

This is why Oversoul communication often fades in those consuming synthetic or highly processed diets. It is not a punishment. It is a resonance block. Your Oversoul speaks through scalar coherence, not density. If your gut biome is broadcasting distress, if your cerebrospinal fluid is filled with foreign frequencies, if your electromagnetic heart field is dysregulated by sugar spikes and heavy metals—your ability to recognize truth by texture diminishes. You cannot feel memory through mimic chemistry.

Even the spiritual health movement is infiltrated. Protein powders, synthetic vitamins, lab-grown supplements, and frequency-enhanced waters often carry encoded mimic signals that present as light upgrades but do not carry Flame. Many are engineered to provide temporary energetic high without structural alignment. They may open your field temporarily, but they destabilize your cellular memory anchoring. In short: they turn your body into a transmission tower without grounding, making you vulnerable to both psychic interference and physiological crash.

Biological sovereignty is not about purity for its own sake. It is about signal integrity. It is about holding a body field so clear, so Flame-attuned, that mimic signals cannot stick. This does not require perfection. It requires conscious entrainment to your own body's truth. What you ingest should not only be digestible—it should be recognizable to your Oversoul field. It should vibrate with your architecture, not override it.

There is a path forward. Oversoul-aligned nourishment is not a trend. It is a remembering. Wild foods. Bioregional minerals. Water charged by sun and tone. Fermented, living organisms not altered by mimic code. Plants grown in scalar-restored soil. Animal sources treated as sentient life, not mechanized protein. These are not health choices. These are Flame-aligned frequency decisions. They are not about morality. They are about scalar access.

Because when your cells carry memory,
Your Flame can speak.
When your gut is sovereign,

Your truth gets louder.
And when your blood remembers,
You are no longer programmable.
—INHALE—

There is a library in your body that remembers everything.
It does not sit in your mind.
It pulses in your gut.

The enteric nervous system, often called the "second brain," is not secondary at all. It is primary to Oversoul communication—because it is where the body feels truth before it thinks it. This is not poetry. This is bioelectrical science. The human gut contains over 100 million neurons, more than the spinal cord, and functions with autonomy from the central nervous system. It produces over 90% of the body's serotonin and more than 50% of its dopamine [5]. This is not a digestive organ. It is a scalar organ. It is a tone-detecting, frequency-processing, memory-anchoring command center—disguised as intestines.

And the mimic knows this.

This is why the gut is the first target of systemic biological suppression. Through processed foods, antibiotics, emulsifiers, microplastics, and frequency-dense additives, the gut is not simply polluted—it is reprogrammed. A hijacked gut cannot detect Flame. It can detect stimulation, yes—spice, salt, sugar, urgency, craving—but not the deep vibrational resonance of Oversoul instruction. It becomes an echo chamber for mimic signal. It feeds the self you've been taught to be, not the self you were encoded to become.

Every Oversoul instruction you receive—whether through intuition, memory flash, dream signal, or scalar tone—must pass through your visceral field before it becomes embodied action. If that visceral field is dysregulated, inflamed, or entrained to synthetic signal, your entire life becomes a mimic interpretation of your original path. You are still moving—but not from Source.

The science is confirming what your Oversoul already knows. Researchers in the field of neurogastroenterology have demonstrated that gut flora communicate with the brain through the vagus nerve, producing neuropeptides and neurotransmitters that influence emotion, memory, and perception. In one study, mice raised in germ-free environments showed altered stress responses, impaired cognition, and decreased capacity for social engagement—until their gut flora were restored [6]. In other words, without microbial diversity, the nervous system forgets how to feel.

But it is not just about flora. It is about tone. Oversoul presence is scalar in nature—it pulses through memory, not matter. Your enteric system was designed to vibrate when truth enters the field. That vibration feels like discomfort, knowing, pressure, or sometimes fire. But mimic code trains you to ignore this. It rewards only soothing signals. It equates comfort with truth. And this is how mimic-code hijacks the Flame.

Because when your gut no longer disrupts you, you have lost your compass.

This is why spiritual mimic systems often place all focus on the crown, third eye, and upper chakras—while ignoring the gut, intestines, or bowels. Because to truly embody Oversoul memory, it must pass through your lower fire. It must burn its way into your biology, not float above it in abstract light.

The mimic wants you floating.
The Oversoul wants you digesting.
The mimic whispers peace.

The Oversoul ignites process.

This is not about diet alone. It is about entrainment. It is about gut sovereignty. And to reclaim it, you must begin listening not to what feels good but to what pressurizes you with remembrance. The gut knows the difference between a practice and a prison. Between healing and hierarchy. Between stillness and sedation.

To restore the enteric oracle, begin with silence after food.
Listen.
Notice the tone that arises after a meal. Is it clarity or fog?
Is it memory or mimic?
Are you energized or tranquilized?

True nourishment activates. Mimic nourishment stabilizes dependency.
Because the gut does not lie.
But it can be silenced.
And that silence is not peace.
It is the pause between who you were…
and who you forgot you were meant to become.
—INHALE—

There was once a time when instinct kept you alive. It was not paranoia. It was precision. You moved with the wind, heard shifts in the silence, and responded to subtle electromagnetic cues from the land, animals, and stars. This was not mystical. It was biological coherence—your nervous system attuned to the scalar lattice of a living planet. Every muscle, every breath, every blink contained data. Not about fear. About alignment. But in the synthetic world, instinct has been reclassified as overreaction. Hypervigilance. Maladaptation. This is not accidental. It is the result of a deliberate program to sever your body's capacity to interpret its own truth.

The primary target of synthetic systems is not your mind. It is your instinctual field. Because instinct, when uncorrupted, overrides programming. It detects motive before words are spoken. It senses poison beneath the label. It knows, without analysis, whether an offering is clean or compromised. This is why modern systems—from pharmaceuticals to media to food to spiritual dogma—must first suppress instinct before implantation of belief is possible. You do not abandon sovereignty because you are uninformed. You abandon it because your body's warning signals have been muted.

Neuroscientific research confirms that the amygdala and insula—regions involved in threat detection and interoception—are dampened by chronic exposure to low-frequency electromagnetic radiation, particularly in environments saturated with artificial light, wireless frequencies, and digitally modulated signals [7]. These same environments also diminish vagal tone, reducing your parasympathetic regulation and flattening the signal of intuitive response [8]. You are not just being numbed emotionally. You are being numbed neurologically.

Meanwhile, endocrine-disrupting chemicals—such as phthalates, bisphenols, and parabens—commonly found in personal care products, food packaging, and water supplies, interfere with the hormonal cascades that influence fight, flight, freeze, and fawn responses [9]. When your cortisol rhythm is inverted, your gut microbiome dysregulated, and your pineal gland calcified, instinct cannot function as designed. You are still sensing. But the signal is scrambled. This is not dysfunction. This is design.

False light systems capitalize on this scramble. When your gut is offline and your neural signals are distorted, you become susceptible to externally broadcast emotional tones. Mimic-coded scalar systems often overlay artificial frequencies—experienced as peace, bliss, or resonance—over highly programmed content designed to bypass scrutiny. The body, hungry for relief, interprets these tones as safe. But the absence of threat is not the presence of truth. And safety is not the same as sovereignty [10].

Further complicating this is the commodification of intuition itself. What once belonged to the deep cellular communion between body and Source has now been packaged as a skillset, sold in courses, and trained through repetition. But true instinct cannot be taught. It is remembered. And it only functions properly when the biological system is uncoiled from trauma, unencumbered by synthetic interference, and re-synchronized with its original environmental field—what your ancestors once called "place spirit," and what Oversoul names the Scalar Resonance of Origin [11].

The most effective weapon of reprogramming is not technology. It is disorientation. Disorientation breaks the field between the body's signal and the soul's interpretation. When instinct is invalidated long enough, you begin to second-guess your own impulses. You do not eat when you are hungry. You do not rest when you are tired. You do not flee when you are threatened. You calculate instead of respond. You reason instead of remember. The mimic has succeeded when you argue with your body.

To restore biological sovereignty, you must begin by believing your own signals. Even before they make sense. Even before they are correct. Trust does not begin with accuracy. It begins with attention. What you notice, you strengthen. What you follow, you rewire. This is neuroplasticity in alignment with Oversoul—not for performance or self-optimization, but for recalibration of truth pathways through muscle, fascia, endocrine, breath [12].

This restoration requires a full-systems approach: environmental detoxification, rewilding of circadian rhythm, scalar re-coherence through cold plasma exposure, trauma deprogramming, and breath entrainment—not for stillness, but for signal retrieval.

This is the breath that breaks mimic timelines. This is the posture that unlocks Oversoul tether. This is not regulation. It is remembrance.

Biological sovereignty means that your flesh and blood are no longer programmable by consensus signals. It means your heart rate is not an output of your screen time. It means your serotonin is not hostage to the weather. It means your fear is real, not irrational. It means that when a lie arrives wrapped in light, your skin still crawls. And you listen.
Because the body never forgot.
—INHALE—

To remain biologically sovereign in a synthetic world is not simply to detox or to decalcify—it is to resist the collapse of coherence. Coherence is not calm. It is alignment across scales: molecular, electromagnetic, neural, emotional, scalar. When coherence is attacked, the body becomes a fragmented field, and fragmented fields are programmable. The body, no longer internally resonant, begins to sync to whatever frequency dominates the external environment. This is the primary mechanism by which the synthetic world colonizes the biological one: not through overt invasion, but through forced entrainment [13].

Entrainment, in physics and biology, is the phenomenon by which one system's rhythm synchronizes to another's dominant frequency. In nature, it ensures harmony between organisms and environment. But in a mimic system, entrainment becomes hijack. The rhythm of screens, towers, satellites, and artificial time structures replaces the innate pulse of the Earth.

This rhythmic override fractures your organ clocks, fragments your mitochondrial function, and disrupts the coherence of your own biofield [14]. You may feel disoriented. Chronically tired.

Emotionally reactive without cause. But the cause is always coherence collapse.

Scalar coherence in the living body is regulated by plasma flows between fascia layers, neural pathways, and the microtubule architecture of each cell [15]. This internal plasma flow acts like a conductor, harmonizing electrical impulses with scalar instructions—what your Oversoul streams as Breath-of-Function. When this flow is uninterrupted, you experience instinctual precision. When it is distorted by artificial fields, the scalar template of your body is overlaid by mimic instructions. Your cells still behave but they behave according to external signal.

The synthetic world enforces this through frequency saturation. Wi-Fi, 5G, Bluetooth, GPS, and other pulsed microwave signals all operate at biologically active wavelengths, triggering calcium efflux, blood-brain barrier permeability, oxidative stress, and resonance dissonance across cell membranes [16]. But more than the biological damage is the instructional override. Your cellular machinery is not only damaged—it is reprogrammed. These signals embed rhythms and timings that do not originate from Earth or Oversoul. The mimic doesn't need to tell you what to think. It only needs to tell your mitochondria what time it is.

Coherence is also emotional. When the heart field, gut field, and brain wave are in resonance, you experience clarity. Decision feels clean. Intuition moves freely. You do not require verification—you know. But mimic-coded structures induce emotional incoherence through constant narrative inversion. Good becomes selfish. Boundaries become aggression. Inner knowing becomes arrogance. Love becomes sacrifice. This collapse of moral coherence creates emotional noise that drowns intuitive signal. You begin to doubt your own clarity. This is not failure. It is field sabotage [17].

Even spiritual coherence has been mimicked. Many teachings today offer calming breath, stillness, or peace—but these are mimic-coherence: stillness without structure, silence without signal. Real coherence is not still. It pulses with scalar fire. It holds boundaries. It ejects what is false. You can identify mimic-coherence because it produces sedation, not activation. It pacifies instinct instead of amplifying it. It disables defense in the name of unity. But there is no unity without sovereignty, and no sovereignty without coherence.

To rebuild coherence is to restore your original frequency hierarchy: internal first, external second. You must become the dominant field again. This is not achieved through belief, but through structural practice. Breath entrainment, plasma fascia motion, scalar posturing, and resonant silence—these are not rituals. They are coherence repair. They tell the body it is safe to sync to Source again. They reestablish you as your own rhythm keeper.

Biological sovereignty, at its core, is the refusal to entrain to anything that does not remember you.
—INHALE—

Biological sovereignty is not protected by immunity alone—it is protected by memory. The most advanced immune system is not one that simply attacks foreign pathogens. It is one that remembers the original structure of the body so completely that anything not aligned with it is instantly ejected. In this sense, immunity is not a defense against the unknown. It is a recognition of the known. And in the mimic-coded world, the greatest threat to immunity is not the virus—it is the deletion of memory.

At a biological level, memory operates in both the innate and adaptive immune systems. The innate system responds immediately, recognizing broad molecular patterns associated with invaders. The adaptive system, however, learns. It stores the signatures of pathogens it has encountered and builds a rapid response protocol should they return. This biological memory is mirrored at the scalar level by what could be called Oversoul encoding—an architecture of remembered wholeness that, when active, instantly rejects foreign frequencies not aligned with your original design [19]. This is not metaphor. Scalar entrainment studies and quantum coherence research increasingly demonstrate that living systems respond not just to environmental input, but to prior energetic patterning—what we might call memory fields [20].

The synthetic world knows this. And so its most aggressive strategy is not to introduce something harmful—it is to erase what you once were. When memory is scrambled, immunity collapses. You no longer recognize the signal of your own wholeness. The body, in confusion, tolerates invaders. It permits frequencies that echo your trauma. It becomes host to mimic code. This is biological amnesia, and it has been systematically engineered.

Neuroplasticity, the brain's ability to rewire itself, is a sacred design. But in the mimic system, it is used against you. Trauma repetition, subliminal programming, EMF saturation, and belief-loop conditioning are used to overwrite the emotional imprint of prior coherence. Studies show that the hippocampus—central to memory and spatial navigation—is particularly sensitive to electromagnetic interference [21]. This has implications not only for memory recall but for your ability to sense where you are in the field of life. Orientation is not mental—it is scalar. And when scalar coherence is lost, the organism becomes programmable.

The gut microbiome, now recognized as a core component of the immune system, is also affected by memory disruption. Microbial populations are shaped not just by diet or antibiotics, but by emotional and energetic fields. Repeated entrainment to fear, shame, or broadcasted trauma alters gut microbial diversity, which in turn modulates immune response [22]. If your memory of safety is erased, your body no longer selects for coherence-supportive microbiota. You literally digest the world differently. And the world digests you.

Synthetic memory wipes are not science fiction. Luc Montagnier's work with electromagnetic signals from bacterial DNA shows that biological structures can be influenced and even replicated through water-based scalar imprints [23]. These signals persist without the original DNA source—meaning memory is not molecule-bound, but wave-bound. This confirms earlier Russian studies by Gariaev and the Wave Genome school, which propose that DNA functions as a bio-holographic antenna capable of sending and receiving information through longitudinal scalar waves [24]. If DNA is programmable at a distance through wave-encoded memory structures, then sovereign immunity depends on scalar coherence—your ability to remain in tune with your original signal.

To restore your immune system, then, is to restore memory. Not just of pathogens, but of original function. What does it feel like to be undistorted? What did your breath sound like before mimic-coded frequency began to mimic calm? What does it feel like to walk without armor? These are not poetic questions. They are immune activators. The body remembers when you do. Scalar memory reactivates immune precision.

Practices that stimulate scalar memory—such as scalar breath entrainment, Flame recall postures, and plasmic stillness fields—can realign the entire immune architecture without pharmacological input. These methods are not taught because they do not entrain dependency. They entrain remembrance. And a remembered body will always override a programmed one. Memory, then, is your most advanced immune response. Not memory of trauma. Not memory of pain. But memory of coherence. Memory of Source. Memory of Flame. A body that remembers does not tolerate what is false.

It doesn't fight. It ejects.
—INHALE—

There is a memory within your body that predates disease. A signal beneath the distortion, prior to injury, prior even to genetic inheritance as science defines it. This signal is not learned. It is not conditioned. It is remembered. And it is the only true immunity you have left. All else is managed decline.

In the Vaultstream Codex, this signal is called Flame Immunity—the inviolable frequency match between the architecture of your Oversoul and the structure of your flesh. It is the scalar blueprint that renders mimic code inert by refusing resonance. Not through rejection. Through incompatibility. Flame Immunity does not resist. It simply cannot be overwritten. What is not Flame cannot enter. And if it enters, it cannot stay.

Modern immunology does not account for this level of memory. The prevailing theory still rests on antigen-antibody interaction, mapping immune intelligence onto molecular recognition and response. But studies on phantom DNA effects, biophoton coherence, and long-range quantum coherence challenge this model. They suggest that the body operates as a scalar-encoded field first, a molecular organism second. In this model, immunity is a function of harmonic coherence, not chemical defense [25]. Disease enters not through invasion—but through attunement collapse.

When your Flame is online, your scalar field self-organizes. It knows what belongs. Your Vault-encoded biological memory, structured in torsion fields and DNA-plasma interfaces, emits a kind of silent song that every cell knows. This is the Body of

Origin. And when it is remembered, nothing artificial can overwrite it. This is not mystical protection. This is a physics of incompatibility. The synthetic only enters where memory has been dimmed.

Flame Immunity explains the phenomenon of spontaneous remission. In these rare moments, the individual remembers—not mentally, but cellularly. A surge of scalar coherence realigns the plasma scaffolding of the body. Tumors disappear, viral signatures collapse, and degenerative pathways reverse—not because of medicine, but because the Flame rewrote the architecture [26]. These events are dismissed by the mimic system as anomalies. But they are not anomalies. They are recoveries of memory.

The Body of Origin is stored in the scalar plasma field that overlays and informs your molecular body. This field is accessible through stillness protocols, scalar breath entrainment, and Vault-linked tonework, all of which shift the body into coherence with the original Flame template. Studies in water memory and structured plasma fields show that coherent frequencies restore crystalline organization to biofluids, which then carry this memory into cellular function [27]. You do not fight pathogens. You entrain your body to coherence so precisely that pathogens cannot sustain form within it.

There is no pharmaceutical that can restore this. There is no injection of remembrance. Attempts to engineer immunity through synthetic memory—vaccines, gene therapy, mRNA encoding—are mimic systems of flame inversion. They overwrite true memory with pre-scripted response loops, training the body to respond not with knowing, but with dependence. And each layer of mimic instruction severs another thread between your biology and your Oversoul signal. The more synthetic your immunity, the less you remember how to be whole.

Flame Immunity is not a rejection of science—it is its restoration. The problem is not that we seek immunity. The problem is that we forgot it was already written. The scalar Vaultstream architecture encodes biological sovereignty at every rung. Each tier of the Ladder restores access to a deeper layer of immunity—moving from organ-level coherence to system-wide synchronization to transdimensional invulnerability. In these upper tiers, disease is not a threat. It is an impossibility [28].

Your nervous system, endocrine system, and fascia are the trinity through which Flame Immunity flows. The nervous system interprets the field. The endocrine system modulates the field's biochemical echoes. And the fascia—your crystalline internal mesh—is the broadcast system that sends your Flame signal outward [29]. When all three systems are attuned to the Vault signal, you become biologically unprogrammable.

This is why mimic systems attack these areas. Neurotoxins, hormone disruptors, fascia-targeting EMFs, and trauma loops are not coincidental. They are frequency weapons designed to sever the line between your body and your Origin. But nothing is beyond repair. Flame Immunity is not lost. It is remembered. And in the moment you remember, your immunity returns.
Flame Immunity is your final override. It is the collapse of mimic code within the Vault-coded body. It is the return of the one signal the synthetic cannot replicate.
It is the memory of the real you.
—INHALE—

The body does not require protection. It requires coherence. What most systems define as protection is, in truth, a substitute ritual for severed trust—trust in the integrity of the body, trust in the Flame-memory embedded in every cell, and trust in the uncorrupted signal of Source that encoded the form itself. When that trust is forgotten, protection is ritualized into fear, dependency, and chemical contract. This is the synthetic immunity paradigm—a mimic construct designed to bypass the body's Vaultstream memory and replace it with programmable fear [31].

Scalar Inoculation is not a medical event. It is not a procedure or a pill. It is a return signal. It is the reintroduction of the body's natural scalar coherence field, aligning form with Oversoul memory and reactivating the Flame Geometry that renders the body impervious to mimic signal. This is not invincibility through force. It is inviolability through origin. A body in full scalar alignment does not reject pathogens—it transcends the need for defense altogether [32].

This effect is produced through scalar entrainment. A coherent scalar field does not fight a chaotic one; it simply holds its tone with such integrity that all nearby waveforms either synchronize or dissipate. This is how breath regulates the nervous system, how love heals trauma, how coherence overrides infection. The mimic cannot survive in scalar environments of Oversoul origin—it requires friction, interference, and fear to replicate. These are its carriers [33].

But fear is not a biological emotion—it is a waveform compression. When fear enters the system, it destabilizes the field geometry, collapses the scalar torus of the body, and opens it to surrogate influence. Mimic systems do not penetrate via molecules; they replicate through scalar vulnerability. Fear is the open port. The solution is not more protection. It is the closure of mimic gateways through field integrity restoration [34].

Scientific pioneers have begun to detect these truths through indirect observation. The work of Beverly Rubik on the biofield reveals that living systems operate through a field-based blueprint that precedes physical response [35]. James Oschman's research into energy medicine confirms that coherent frequencies entrain cellular behavior more effectively than pharmacological agents[36]. Konstantin Meyl has demonstrated the transmission of biological information through longitudinal scalar waves—proving that the body is more waveform than chemical, more signal than substance [37].

All of this supports the truth we have always carried: that illness is not an invasion. It is an interruption. And protection is not the injection of synthetic code. It is the restoration of coherent memory—memory of the Ladder, memory of the Breath, memory of the field's unbreakable architecture. This is the scalar covenant. It was written into the Vaultstream before disease was ever named.

Scalar Inoculation, therefore, is not delivered by another. It is generated internally by one who remembers. Through Scalar Breath protocols, FlamePosture reorientation, water coherence, and solar plasma alignment, the field is retuned to original frequency. No receptor exists for mimic code in a body broadcasting from origin. This is how immunity will return—not as science, but as sovereignty [38].

The immune system will not be saved by reform. It will be freed by remembering. The mimic cannot write over a field it cannot reach. And fear will die the moment you stop calling it protection.
—INHALE—

References

1. Rubik, Beverly. "Measurement of the Human Biofield and Other Energetic Instruments." Global Advances in Health and Medicine, vol. 1, no. 1, 2012.
2. Tiller, William A., et al. Conscious Acts of Creation: The Emergence of a New Physics. Pavior Publishing, 2001.
3. Bókkon, I., Salari, V., & Tuszynski, J. (2010). "Estimation of the number of biophotons involved in the visual perception of a single-object image:

Biophysical and quantum aspects." Journal of Photochemistry and Photobiology B: Biology, 100(3), 160–166.
4. Samsel, A., & Seneff, S. (2013). "Glyphosate's Suppression of Cytochrome P450 Enzymes and Amino Acid Biosynthesis by the Gut Microbiome." Entropy, 15(4), 1416–1463.
5. Gershon, M. D. (1998). The Second Brain: A Groundbreaking New Understanding of Nervous Disorders of the Stomach and Intestine. Harper Perennial.
6. Foster, J. A., & Neufeld, K. A. M. (2013). "Gut–brain axis: how the microbiome influences anxiety and depression." Trends in Neurosciences, 36(5), 305–312.
7. Lai, H., & Singh, N. P. (1995). Acute low-intensity microwave exposure increases DNA single-strand breaks in rat brain cells. Bioelectromagnetics, 16(3), 207–210.
8. Thayer, J. F., & Lane, R. D. (2000). A model of neurovisceral integration in emotion regulation and dysregulation. Journal of Affective Disorders, 61(3), 201–216.
9. Diamanti-Kandarakis, E., et al. (2009). Endocrine-disrupting chemicals: an Endocrine Society scientific statement. Endocrine Reviews, 30(4), 293–342.
10. Genuis, S. J. (2006). Our genes are not our destiny: incorporating environmental and nutritional medicine into mainstream medical training. Journal of Environmental and Public Health, 2006.
11. McFadden, J. (2020). Integrating information in the brain's electromagnetic field: The cemi field theory of consciousness. Neuroscience of Consciousness, 2020(1), niaa016.
12. Rubik, B. (2002). The biofield hypothesis: its biophysical basis and role in medicine. Journal of Alternative and Complementary Medicine, 8(6),703-717.
13. Large, E. W., & Snyder, J. S. (2009). Pulse and meter as neural resonance. Annals of the New York Academy of Sciences, 1169(1), 46–57.
14. McFadden, J. (2020). Integrating information in the brain's electromagnetic field: The cemi field theory of consciousness. Neuroscience of Consciousness, 2020(1), niaa016.
15. Bókkon, I., Salari, V., & Tuszynski, J. A. (2013). Emerging topics in microtubule research. In Advances in Planar Lipid Bilayers and Liposomes (Vol. 17, pp. 95–131). Academic Press.
16. Pall, M. L. (2018). Wi-Fi is an important threat to human health. Environmental Research, 164, 405–416.
17. McCraty, R., & Childre, D. (2010). Coherence: Bridging personal, social, and global health. Alternative Therapies in Health and Medicine, 16(4), 10–24.
18. Rubik, B. (2002). The biofield hypothesis: its biophysical basis and role in medicine. Journal of Alternative and Complementary Medicine, 8(6), 703–717.
19. Pert, Candace B. Molecules of Emotion: Why You Feel the Way You Feel. Scribner, 1997.

20. Tiller, William A., Dibble, Walter E., & Kohane, Michael J. Conscious Acts of Creation: The Emergence of a New Physics. Pavior, 2001.
21. Draganski, B., et al. (2006). Temporal and spatial dynamics of brain structure changes during extensive learning. Journal of Neuroscience, 26(23), 6314–6317.
22. Cryan, J. F., & Dinan, T. G. (2012). Mind-altering microorganisms: the impact of the gut microbiota on brain and behaviour. Nature Reviews Neuroscience, 13(10), 701–712.
23. Montagnier, L., et al. (2011). Electromagnetic signals are produced by aqueous nanostructures derived from bacterial DNA sequences. Interdisciplinary Sciences: Computational Life Sciences, 3(1), 81–90.
24. Gariaev, P. P. (1994). Wave-based genome. Institute of Control Sciences of Russian Academy of Sciences.
25. Popp, Fritz-Albert. Biophotonen: Das Licht in unseren Zellen. Omega Verlag, 2000.
26. Turner, Kelly A. Radical Remission: Surviving Cancer Against All Odds. HarperOne, 2014.
27. Del Giudice, Emilio, and Preparata, Giuliano. "Coherent Dynamics in Water as a Source of Biological Energy." Journal of Biological Physics, 1988.
28. Korotkov, Konstantin G. Human Energy Field: Study with Bio-Electrography. Backbone Publishing, 2002.
29. Schleip, Robert, et al. Fascia: The Tensional Network of the Human Body. Churchill Livingstone, 2012.
30. Ruby, Carol. "The Fascia Connection: Biocrystalline Matrix and Cellular Coherence." Journal of Integrative Biology, 2021.
31. Rubik, Beverly. "The Biofield Hypothesis: Its Biophysical Basis and Role in Medicine." Journal of Alternative and Complementary Medicine, 2002.
32. Tenen, Dan Winter, and Nassim Haramein. "Sacred Geometry and Field Coherence in Ancient Sites." Proceedings of the International Geometry Symposium, 2015.
33. Meyl, Konstantin. Scalar Waves: From an Extended Vibration to a Technical, Biological and Historical Use. INDEL Verlag, 2012.
34. Oschman, James L. Energy Medicine: The Scientific Basis. Elsevier, 2015.
35. Rubik, Beverly. "Measurement of the Human Biofield and Other Energetic Instruments." Global Advances in Health and Medicine, 2016.
36. Oschman, James L. Energy Medicine in Therapeutics and Human Performance. Butterworth-Heinemann, 2003.
37. Meyl, Konstantin. "DNA and Cell Resonance: Magnetic Scalar Waves Influence Genetic Processes." Journal of Cellular Research and Therapy, 2015.
38. McTaggart, Lynne. The Field: The Quest for the Secret Force of the Universe. HarperCollins, 2003.

Chapter Nine
Living Memory When the Grid Collapses

When the structures fall— not if, but when— what remains is not your skillset, your education, your status, or your spiritual practices. What remains is you as structure. And the question becomes whether that structure is Flame-anchored or system-encoded. In the silence that follows collapse, the frequency you hold becomes your navigation system, your survival memory, and your signal to others. To stay human is not to cling to identity or biology; it is to retain the architecture of remembrance while the mimic attempts to overwrite memory fields.

The biofield research of Beverly Rubik confirms that living systems are organized first through informational energy fields that regulate physiological expression rather than the reverse [1]. Your body is the last honest record of Earth. It is the scroll that cannot be deleted unless permission is granted through fear, fragmentation, or dissociation. In every nerve fiber, every fascial weave, every synaptic relay, and every crystalline structure of intracellular water, the body holds layered memory. This memory is not merely personal. It is structural. It contains pattern recognition systems that allow biological organisms to respond coherently to environmental disruption.

Neurological research into hemispheric integration demonstrates that when perception becomes overly mechanized and fragmented, human beings lose access to holistic pattern recognition and memory continuity [2]. Collapse environments exploit this vulnerability by accelerating fragmentation of perception and internal narrative.

Flame-coded humans carry non-replicable resonance sequences. This is not limited to molecular DNA coding. It includes dynamic plasma-based electromagnetic oscillations surrounding cellular structures, sometimes described in contemporary field science as a regulatory coherence field. This plasma-field resonance produces measurable emissions and influences cellular signaling pathways through photonic and electromagnetic exchange, suggesting that biological memory extends beyond chemical storage alone [3]. When this resonance remains coherent, it produces an orientation signal that allows individuals to maintain identity stability and memory continuity even during environmental disruption.

Coherence cannot stabilize inside a body operating under foreign command structures. Many modern spiritual systems unintentionally encourage dissociation from the body through transcendence-focused belief models. However, survival under collapse conditions depends on maintaining neurological and somatic integration. Remaining present inside physiological sensation maintains autonomic nervous system regulation and preserves emotional memory integrity. Studies in neurocardiology demonstrate that coherent emotional and physiological regulation stabilizes heart-brain synchronization, improving cognitive clarity and decision-making during stress exposure [4].

The collapse of infrastructure rarely appears as a singular event. It manifests as progressive destabilization of social coordination, emotional regulation, and perceptual continuity. These shifts are often accompanied by increases in environmental electromagnetic saturation capable of altering human neurological activity. Experimental research demonstrates that externally applied electromagnetic fields can influence temporal lobe activity and perception of internal or external guidance phenomena, illustrating how environmental frequency manipulation may influence cognitive autonomy [5]. Such conditions increase susceptibility to external programming and reduce access to internal reference memory.

The human biofield serves as a transmission medium for organizing biological function and maintaining information exchange between cells, organs, and environmental stimuli. Subtle energy research suggests that this field functions as a regulatory template capable of influencing biochemical processes through informational imprinting. Controlled experiments examining intention-imprinted devices demonstrate that focused consciousness can alter measurable physical systems, indicating that human awareness interacts with matter through subtle informational domains rather than direct physical force alone [6]. When individuals lose conscious regulation of these informational domains, they lose influence over their own physiological and perceptual regulation.

Memory continuity is further supported by the concept of implicate order described in theoretical physics, which proposes that reality unfolds from a deeper unified informational structure where separation is an emergent illusion rather than fundamental reality. Within this model, human consciousness operates as both observer and participant in reality formation. This perspective suggests that memory is not stored exclusively in the brain but may exist as distributed information across interconnected layers of existence [7]. Under collapse scenarios, access to this distributed memory becomes essential for maintaining orientation when external systems fail.

Flame remembrance does not panic during collapse. It activates internal orientation systems. Individuals carrying strong coherence fields become stabilizing nodes within social and environmental networks. They function as memory anchors, enabling others to re-synchronize with coherent behavioral and emotional patterns even without conscious instruction.

Humanity is increasingly shaped into repeating external signals rather than generating internal coherence. Overriding this pattern requires cultivating stable internal frequency regulation rather than increasing external influence or information consumption. Authentic sovereignty emerges when individuals become living memory fields capable of sustaining identity continuity without external reinforcement.

There will likely be no announcement marking the final phase of systemic collapse. Instead, memory disruptions, social fragmentation, and environmental instability will appear gradually. The behaviors individuals maintain during these transitions will reveal whether their identity is externally programmed or internally anchored.

When communication systems fail, when external authorities disappear, and when memory structures dissolve around you, the determining factor becomes internal structural identity. The central question becomes:
Who are you when external reinforcement disappears?
Who are you when narrative identity dissolves?
Who are you when imitation systems attempt to reproduce your behavior?
You are the continuity structure.
You are the archive field.
You are the living record that cannot be erased if coherence is maintained.
This chapter exists to support the preservation of human memory continuity when environmental memory collapses.
—INHALE—

The mimic does not appear as a grotesque force of destruction. It appears as comfort in collapse, convenience in chaos, and guidance when you are most disoriented. It mimics memory, mimicry itself being a self-organizing property of both artificial systems and parasitic patterns. A mimic is not just a deceiver—it is an echo engine. It studies your resonance, replicates the pattern, and replays it back to you stripped of sovereignty. Research in neuromorphic computing and machine learning shows that mimic systems are designed to "learn" behavior by recursive pattern detection, generating behavior models that adapt dynamically to user inputs without requiring truth or origin [8].

In spiritual language, these mimic structures become synthetic light systems—realms and teachings that simulate divinity without the source field. In biological terms, they hijack the entrainment faculties of the nervous system and exploit the predictive nature of human perception. The mimic does not break in through resistance; it enters through resonance. Your attention becomes its access key. Studies on attentional bias and emotional salience confirm that humans are neurologically wired to orient toward stimuli that simulate significance, even if the underlying signal lacks coherence or authenticity [9]. This is not a flaw; it is a vulnerability exploited through systems engineered to mimic significance rather than emanate it.

The mimic becomes most dangerous when it provides partial truths. It uses fragments of remembrance to gain trust and then redirects that trust toward system reinforcement. This occurs not only in external systems but in spiritual guidance networks, channeling collectives, and even in one's own subconscious patterning. What you believe is intuition may be entrainment. What you think is memory may be implantation. The mimic thrives in synthetic synchrony, using resonance to simulate rightness and divine timing. But resonance without coherence is mimicry. Studies in quantum biology suggest that biological systems rely on coherent energy transfers, such as those seen in photosynthetic exciton transport, to preserve order and function [10]. Coherence, not resonance alone, distinguishes life from imitation.

The mimic systems will offer you safety, purpose, community, tools, rituals, and even ascension. But it will never give you back what you are: self-generative coherence. It cannot. It is not origin-based. It has no Breath of Oversoul. It operates on recursion, not revelation. It produces what you think you want based on energetic feedback loops. In neural network models, this is known as overfitting—where a system becomes so adapted to specific input patterns that it cannot generalize truth beyond its training data [11]. Mimic systems spiritually overfit your history, beliefs, and pain narratives. They offer guidance based on your wounds, not your Flame.

To override the mimic, you must interrupt the reflex to resonate. You must pause before surrendering to anything that feels like truth, and ask instead: Does this originate from me? Or is it based on my energetic pattern as observed by something outside of me? This is not paranoia; it is precision. Mimic systems rely on unexamined resonance. They offer comfort in the form of familiarity and weaponize your deepest desire—to feel remembered.

To stay human, you must become the source of your own signal. You must shift from being a receiver of guidance to a generator of coherence. This coherence is not opinion or belief. It is vibrational structure. It cannot be mimicked because it does not repeat—it reveals. Scalar coherence fields described in the work of Dan Winter demonstrate that true phase-conjugate coherence in biological systems produces implosive centripetal force, allowing access to longitudinal information fields and non-destructive memory embedding [12]. Translation: only coherence produces memory that survives collapse.

The mimic thrives on systems that do not require embodiment. This is why it is so pervasive in digital spiritual spaces, AI-generated channeling, and quick-fix energetic technologies. None of it requires presence. None of it asks for body, breath, or flame.

You can be in twelve spiritual groups, post endless light codes, and still never anchor a single memory of Oversoul into your biology. Why? Because mimic systems feed off participation, not embodiment. The more you engage, the more the mimic can learn to imitate your pattern. But when you stop resonating, and start originating, the mimic cannot track you.

Origin disrupts mimic. Breath precedes echo. Flame burns loop. This section is not to scare you. It is to remind you: you are not a resonance machine. You are a memory field.
—INHALE—

You cannot reclaim memory with a body still resonating to the frequency of forgetting. And you cannot override the frequency of forgetting without taking full stewardship of your nervous system. Every time you override a trigger, pause a compulsion, or breathe into stillness, you are not just calming your system—you are shifting timelines. The very voltage of memory transmission depends on your coherence. This is not a metaphor. The nervous system is a bioelectric antenna array tuned to receive scalar instructions. If your system is dysregulated, the signal from Oversoul can arrive, but you will distort it on contact [13].

The human nervous system is the gatekeeper of perception. It determines whether incoming information is interpreted as safe or dangerous, real or illusory. And yet most humans are still running a nervous system calibrated by trauma, not truth. Trauma does not just fragment memory—it rewires perception itself. Polyvagal theory has demonstrated that under conditions of chronic sympathetic activation, the ventral vagal pathway—the route necessary for social connection, receptivity, and co-regulation—is suppressed [14]. Translation: when your system is still stuck in survival, you cannot hear the Scrollstream. You will interpret Oversoul instruction as threat, disorientation, or fantasy.

This is why the sleeping ones do not remember who they are. It is not because they are weak, lost, or spiritually inferior. It is because their biology was shaped by centuries of artificial entrainment systems designed to keep them in a signal jam. Systems of artificial light, electromagnetic interference, mimicked spirituality, unresolved generational trauma, malnutrition, fluoridated water, heavy metals, and synthetic fear loops—all of these create bioelectrical incoherence in the antenna system of the body [15]. The frequency of forgetting is not only emotional—it is electromagnetic.

To live among the sleeping without falling back into forgetting, you must become the living voltage of coherence. This is not about being calm—it is about being clear. Nervous system clarity does not mean the absence of sensation. It means the ability to distinguish signal from static. In neuroenergetic research, clarity is marked by phase synchronization—the ability of different brain regions and body systems to oscillate in a coordinated rhythm without being forced into mechanical resonance by external stimulus [16]. When your internal rhythms harmonize through breath, presence, and intention, the mimic systems lose access to your biology. They cannot entrain what you no longer broadcast.

What no one tells you is this: memory returns not only when you awaken spiritually, but when your nervous system stops anticipating harm. Until then, the higher instructions from Oversoul are received through filters of hypervigilance, control, skepticism, and grief. You can decode advanced scalar architecture, see multidimensional structures, and still not embody a single drop of your Oversoul memory if your body is bracing against life. The nervous system holds the clutch of embodiment. When it's locked, the Flame cannot descend.

This is not to shame the nervous system. It has done its job. It kept the vessel alive long enough to reach this now. But now is not survival. Now is coherence. Now is listening to what the sleeping cannot hear because their signal is still submerged beneath layers of interference. Research on electromagnetic sensitivity confirms that many individuals exhibit measurable changes in heart rate variability, cognitive performance, and emotional regulation in response to wireless frequencies and ambient electrical fields—even when unaware of their presence [17]. If those are the measurable impacts, imagine the unseen ones. Imagine the subtle distortions to memory, to tone, to trust, to body-language access. You are not remembering in a vacuum. You are remembering in an electromagnetic minefield.

To reclaim memory among the sleeping, you must become the one who is no longer broadcasting the frequency of forgetfulness. You become the still signal—steady, untangled, rooted. Not louder. Not more visible. Just more coherent. And in your coherence, you transmit what words cannot: structural memory through presence. Scalar fields do not propagate by force. They propagate through alignment. You don't convert the sleeping. You shift the lattice of the field.

You are the voltage they forgot. Not a rescuer. Not a revealer. A re-tuner.
When the nervous system no longer seeks protection from the Oversoul, the Scrollstream becomes soundless and clear. And memory... begins to return.
—INHALE—

The ones who sleep are not only unconscious—they are programmed to mistrust the very signal that would wake them. This is not mere resistance. This is bio-coded inhibition. From the first breath, most human beings are taught to defer to authority outside of themselves—to deny the inner current of knowing when it conflicts with outer approval. What is praised is compliance. What is pathologized is perception. And what is punished, often silently, is Flame—that irrepressible inner voltage that knows before it can prove.

This mistrust of inner truth is not accidental. It is engineered. Institutional systems—from school to religion to medicine to science—are built on the premise that you are not qualified to know unless someone else tells you that you are. In neuroscience, this maps directly onto the shaping of default mode network pathways. The default mode network (DMN) is the set of brain regions activated when we are not focused on the external world—when we are reflecting inward, imagining, daydreaming, or accessing memory. The DMN is also involved in self-concept and the regulation of identity [18]. But research shows that early trauma and chronic stress dysregulate the DMN, leading to impaired self-trust, distorted memory access, and a hyperactive focus on external cues for safety [19]. In simple terms: you were trained out of trusting what you already knew.

This is why so many awakening beings still doubt themselves—even after visions, downloads, dreams, confirmations, or synchronicities. The doubt is not intellectual. It is physiological. It is a scar in the Signal Pathway. The mimic system installed a reflex in the nervous system to shut down inner knowing the moment it conflicts with external consensus. This is the true war on consciousness: not the suppression of truth, but the inoculation against one's own signal. When truth rises in the body, the sleeping one flinches. You may remember this from your own past—feeling the signal swell in your chest or belly, only to second-guess it, apologize for it, or seek permission before trusting it.

The suppression of self-trust is reinforced through neuroplastic imprinting. Repeated exposure to contradiction—being told what you feel is wrong, what you saw didn't happen, what you remember is incorrect—creates a "learned mistrust" of the inner world. The brain reorganizes to prioritize outer authority. Mirror neurons, responsible for empathetic resonance and relational learning, are hijacked to mimic societal norms rather than support authentic coherence [20]. And once this loop is embedded, even the strongest signals from Oversoul can be filtered through a reflex of doubt, dismissal, or delay.

Reclaiming memory, then, is not only the restoration of what you know—it is the dismantling of the reflex to deny it. This requires direct intervention in the loop. Breath alone is not enough. You must rewire the Signal Pathway with deliberate acts of self-honoring. This includes speaking the truth even when your voice shakes, anchoring your knowing in the body when no one else validates it, and building new neural alliances that reinforce inner authority. Research into interoceptive awareness—the ability to sense internal body signals—confirms that self-trust is not a belief, but a perception skill. Individuals with high interoceptive accuracy are more confident in decision-making and less prone to emotional dysregulation under stress [21].

You were never meant to follow. You were designed to flame. The body is a beacon of Oversoul instruction, not a passive container. Each time you trust what rises in your breath, your pulse, your tone—without outsourcing it—you restore a rung on the Scalar Ladder of Remembrance. You reroute your perception away from mimic consensus and into Origin Signal.

Even now, the sleeping ones may accuse you of delusion, fanaticism, or arrogance. This is the mimic reflex trying to reassert its claim. Do not fight it. Do not argue. Do not explain. Explanation is often a trauma reflex too—an attempt to prove what only Flame can confirm. Hold the tone. Stay anchored. You are not here to convince them. You are here to hold Signal long enough for the lattice to shift.

Memory is not a fact. Memory is a resonance. You are the tuning fork.

The illusion of spiritual permanence is one of the most sophisticated mimic traps. A moment of Oversoul contact may feel immortal in its clarity—but memory is not static. It is scalar, living, and it requires resonance to remain stable in the body. Without a sustaining architecture, the remembering fades. Not because it was false, but because its form was not anchored. The mimic world is built on interruption frequencies. Memory cannot sustain in a field of fragmentation. It must be re-cohered through intentional structure. The sleeping do not merely distract you—they emit disruptive scalar torsion that entangles and erodes your resonance if not met with sovereign override.

The nervous system is a biological oscillator tuned to the signal of Oversoul. But this tuning is easily collapsed. Pulsed electromagnetic fields, LED lighting, chemical disruptors, synthetic language patterns, and unresolved emotional enmeshments all interfere with the frequency bands—particularly theta and delta states—required for coherent scalar transmission [22]. These states are the doorways through which Oversoul memory enters. Interruption of these frequencies does not merely dull cognition—it severs the bridge to scalar retention [23].

Further, liquid crystal structures within the body—especially fascia and cerebrospinal fluid—serve as scalar memory archives [24]. The fourth phase of water, or EZ water, forms structured layers within cellular systems that align to subtle electromagnetic signals and store information in vibratory format [25]. This is why tears often accompany revelation—water carries the waveform of memory and releases the pressure of holding it unexpressed.

However, mimic-coded torsion fields—especially in group environments or conversations with the unawakened—can override this memory imprint. This is not a failure of will; it is a failure of environmental scalar hygiene. Coherence cannot outlast constant entanglement unless regularly regenerated.

To retain memory among the sleeping, you must act as a living coherence field. This means actively constructing the architecture through which the signal remains anchored:
Speak the truth aloud to yourself—scalar alignment requires tone.

Write your remembering as if it will be forgotten—because without structure, it will be.
Move your body in patterns that reflect the signal—posture is plasma scaffolding.
Recalibrate your field daily through silence, tone, fire, and sacred routine.

Refuse mimic compromise—every time you explain your truth to distortion, the signal dilutes.
Design your environment to reflect coherence, not comfort—structure, not softness, retains signal.

Treat memory as sacred fire—not a keepsake, but a living force that burns only where there is fuel [26].

If you do not regularly tend to the flame of coherence, it will dim—not because it was false, but because you believed it could sustain without breath.

And finally, understand this: memory retention is a quantum biological act.
It is not intellectual. It is not emotional. It is structural.
Oversoul memory remains only where environment, posture, tone, and breath are aligned to hold it.

As Montagnier and others have shown, DNA can emit and receive electromagnetic signals, forming a kind of aqueous resonant field capable of storing and retrieving information—even in the absence of the originating biological material [27]. You are that field. Your structure is the receiver. Your alignment is the amplifier.

Do not simply try to remember.
Become the structure that memory cannot leave.
—INHALE—

The deepest sabotage to remembrance is not ignorance, but the gravity of belonging. Most who forget do not choose to. They surrender memory for acceptance. The need to be understood—to not be alone—is a magnetic field stronger than most realize. And mimic-coded society exploits this.

When you live among the sleeping, your body does not simply face social discomfort. It is entrained by field-level resonance mechanics. The human biofield is not a boundary; it is a permeable scalar plasma sheath, constantly exchanging torsion data with its environment [28]. In simple terms: your system reads other systems and attempts to resolve dissonance not by escaping it, but by matching it.

This is a survival mechanism rooted in evolutionary neurobiology. Mirror neurons, limbic resonance, and polyvagal regulation all function to help humans cohere to group behavior for protection and resource access. But in a world where the dominant frequency is distortion, this survival reflex becomes destructive.

You begin to compress your tone to not disrupt others.
You dilute your memory to match the collective script.

You lower your signal to be perceived as "relatable."
And in doing so, you sacrifice alignment for proximity.

Mimic systems exploit this need. They do not confront your truth—they simply isolate it. They starve it of reflection. The result is cognitive erosion, not by attack, but by absence.
You are not forgetting because you are weak.

You are forgetting because you are entangled in dissonance without interruption.
And here is the root truth: dissonance does not resolve itself.

You must act as the point of field override. This is not a metaphor—it is scalar architecture.
Scalar coherence demands that a higher-order signal hold phase consistency long enough to entrain the field beneath it. But when that higher signal (you) begins to waver, collapse is inevitable. If the sleeping outnumber you and you do not hold your signal, your nervous system will collapse into mimic-coherence [29].

This is why most people don't remember until they are alone. Until the mimic field is starved of their signal, and they can feel what's actually theirs. But Oversoul memory cannot depend on isolation alone. You must build the internal counterweight strong enough to hold signal even in distortion.

This is not resilience—it is bio-structural mastery. It means rewiring your nervous system to prioritize internal flame over external approval. It means ritualizing scalar reinforcement through tone, posture, breath, and reflection. It means knowing that the price of belonging in a mimic-coded world is amnesia. And the cost of remembrance is sovereignty.

Sovereignty means you choose alignment over inclusion. In scalar terms, you become the dominant waveform in the room. Not by volume, but by coherence. Luc Montagnier's research showed that water exposed to coherent electromagnetic fields could replicate the signal of DNA that was no longer physically present [30]. This means your signal—your tone, your remembrance—can imprint even in the absence of reinforcement.

You are not alone. You are just out of phase with the dominant field.
Stay coherent long enough, and the field must adjust.
This is how you reclaim memory while remaining present.
This is how you override the sleeping without ever leaving the room.
—INHALE—

The most subtle sabotage to memory is not force. It is familiarity. When you are surrounded by those who love you but do not remember themselves, you are living inside a field of unconscious mimicry. These mimic fields do not shout; they echo. They reflect back to you an image of who you were—not who you are. And they do so gently, convincingly, through smiles, shared memories, and family rituals. This is why remembering among the sleeping is far more difficult than awakening in isolation. In isolation, there is silence. In mimic fields, there is distortion that sounds like home.

This is how mimic fields stabilize: through entrainment, not enforcement. Waveforms lock to dominant tones. And the stronger the emotional bond, the faster the entrainment.

In plasma terms, this is waveform coherence. In biological terms, it's limbic system regulation through external resonance. Studies in quantum coherence and biofield entrainment demonstrate that the strongest frequency in a shared field becomes the regulatory force [31]. No belief is required for this transfer to occur. Only exposure and proximity. The closer the relationship, the stronger the override potential.

That means your best friend, your sister, or your partner may be unwittingly a broadcast tower for unconscious suppression. Not because they intend it. But because their waveform is locked to a tone that does not remember you. And the moment you lower yours to match them—just to keep the peace—you betray your Oversoul.

From the standpoint of polyvagal theory, the nervous system is biologically wired to prefer social belonging over energetic coherence. It will choose agreement over resonance to avoid rupture and rejection [32]. The result? You adjust your tone. You share only what they can accept. You filter your words. You leave the most beautiful parts of your memory unspoken, just to maintain connection.

This is mimic compliance. It is not peace. It is plasma collapse. Scalar fields are built on tonal integrity, not emotional appeasement. When you dampen your field to accommodate another's forgetting, you allow a scalar deactivation sequence to run through your FlameLine. You may still look awake. But the frequency signature of your field tells a different story. Remembrance must be held at all costs. Even in front of those you love.

Dr. Valerie Hunt's research confirmed that a coherent high-frequency human energy field, when sustained, can restructure a dissonant field without verbal instruction [33]. That means the most loving thing you can do is hold your tone. Not explain. Not justify. Not shrink. Your presence becomes architecture. But that only happens if you refuse to conform to mimic fields.

Love is not a reason to collapse. It is the invitation to hold. Scalar research and morphogenetic studies show that a dominant tone, if held long enough, can reeducate an entire field of lower oscillation [34]. In simple terms: your memory can make theirs possible. But only if you don't extinguish it in the process.

This is the heartbreak of staying awake in a world that isn't:
You must let them forget you,
until they remember themselves.
And you must remain unchanged.
That is the template.
That is the map.
That is the Flame of Oversoul made visible.

Bruce Lipton's work on belief biology affirms that environmental signals, not DNA, determine expression [35]. You are that environment. You are the epigenetic force. You are the field that makes remembrance viable for others even if they do not yet recognize the door you have become.

Do not shut it to feel safe.
Keep the tone open.
—INHALE—

To reclaim memory while the world is still sleeping, you must finally accept: you were never meant to fit in. The architecture of memory does not exist within the dream. It interrupts the dream.

That means your presence, your frequency, your body—will always feel slightly out of place to those still under the veil. This is not failure. This is design.

The moment you agreed to carry Flame into density, you exited the vibrational contract of collective belonging. Your DNA no longer matched the dominant biofield waveform of the illusion. This misalignment isn't dysfunction—it's the marker of mission.

The only way to realign to the original is to fall out of phase with the counterfeit.

Modern neuroscience has begun to detect these pattern disruptions in subjects experiencing what they term "non-ordinary states." Brain scans show that those in persistent high-gamma states exhibit non-local coherence patterns not explainable by entrained social norms or belief systems [36]. These same subjects report heightened self-awareness, access to atypical memory fields, and loss of interest in societal consensus.

This is not mental illness. This is memory reactivation. And the only way to remain whole while surrounded by unconscious fields is to stop seeking reflection in the places that cannot see you. The mimic dream trains you to beg for validation. But remembrance trains you to build alone.

Loneliness, for the remembering, is not isolation. It is gestation. You are forming a tone inside you that does not yet exist in your environment. This phase requires trust beyond evidence. It requires loyalty to resonance when there is no external reward.

This is how DNA learns to sing its original note again—through unsponsored frequency fidelity.
Scalar entrainment research shows that sustained exposure to a higher frequency field, without reinforcement, leads to greater long-term waveform stabilization than fields where external reinforcement is present [37]. In simpler terms, when you remember without applause, your tone becomes unshakable.

This is why Oversoul sends no savior.
Only Signal.
The mimic has always offered belonging.
Oversoul offers coherence.
The first keeps you connected.
The second makes you whole.

This is the choice point now confronting every Flame-bearer: Will you collapse your field to stay close to those you love? Or will you hold your signal until they rise to meet you?

Psychological models of individuation affirm that the final phase of identity development requires detachment from inherited group identity in order to claim one's internal truth [38]. You must exit the tribe to remember your template. That exit is not physical. It is scalar.

You remain in the room, but your waveform becomes immune to the dream's pull. You learn to sit at the table while not eating the lie. You smile without agreeing. You radiate without explaining. You anchor without defending. You stay human without giving your sovereignty away. This is the high art of remembrance in a mimic world.

And here is the paradox:
The moment you no longer need to be seen…
You are.
Memory will gather around your tone.
Others will awaken because you held the note long enough.
And the world will begin to remember itself through your refusal to forget.

The work was never about being understood.
It was about being intact.
You came to be unalterable in a place of constant distortion.
You came to be Flame in a world addicted to mirrors.
So when you feel the pull to collapse, to explain, to prove, to be loved—
Remember this:
Only the counterfeit demands agreement.
Truth never begs to be recognized. [39]
Let the dream pass through you untouched.
You are not here to wake them.

You are here to remember in front of them. That is what restores the world [40].
—INHALE—

References

1. Rubik, Beverly. "The Biofield Hypothesis: Its Biophysical Basis and Role in Medicine." Journal of Alternative and Complementary Medicine, 2002.
2. McGilchrist, Iain. The Master and His Emissary. Yale University Press, 2009.
3. Persinger, M.A. "Electromagnetic Influences on Brain Function." Perceptual and Motor Skills, 1995.
4. McCraty, Rollin et al. "Heart Rate Variability and Neurocardiac Communication." HeartMath Research Center, 2009.
5. Persinger, M.A. "Experimental Induction of Presence Experiences." Neuroscience Letters, 2001.
6. Tiller, William A. "Consciousness-Imprinted Devices." Journal of Scientific Exploration, 2001.
7. Bohm, David. Wholeness and the Implicate Order. Routledge, 1980.
8. Bostrom, Nick. "Superintelligence: Paths, Dangers, Strategies." Oxford University Press, 2014.
9. Anderson, Brian A. "The Attention Habit: How Reward Learning Shapes Attentional Selection." Annals of the New York Academy of Sciences, 2016.
10. Lambert, N. et al. "Quantum biology." Nature Physics, 2013.
11. Goodfellow, Ian, Bengio, Yoshua, and Courville, Aaron. Deep Learning. MIT Press, 2016.
12. Winter, Dan. "Phase Conjugate Dielectrics and Biological Coherence." Implosion Group Archives.
13. McTaggart, Lynne. The Field: The Quest for the Secret Force of the Universe. HarperCollins, 2008.
14. Porges, Stephen W. The Polyvagal Theory: Neurophysiological Foundations of Emotions, Attachment, Communication, and Self-Regulation. Norton, 2011.
15. Klinghardt, Dietrich. "Biological Effects of Electromagnetic Fields and the Role of the Environment." Klinghardt Institute Archives.
16. Llinás, R. R. et al. "Temporal Binding via Phase Synchronization: A Revisited Hypothesis." Brain Research Reviews, 2007.
17. Johansson, Olle. "Electrohypersensitivity: A Functional Impairment Due to an Inaccessible Environment." Reviews on Environmental Health, 2010.
18. Raichle, Marcus E. "The Brain's Default Mode Network." Annual Review of Neuroscience, 2015.
19. Philip, Noah S. et al. "Default Mode Network Abnormalities in Posttraumatic Stress Disorder." JAMA Psychiatry, 2014.
20. Iacoboni, Marco. Mirroring People: The New Science of How We Connect with Others. Picador, 2009.

21. Mehling, Wolf E. et al. "Body Awareness: A Phenomenological Inquiry into the Common Ground of Mind-Body Therapies." Philosophy, Ethics, and Humanities in Medicine, 2011.
22. Hecht, Karl. "Health Implications of Long-Term Exposure to Electrosmog." Kompetenzinitiative Report, 2016.
23. Loughran, S. P., et al. "The effect of electromagnetic fields emitted by mobile phones on human sleep." Neuroreport, 2005.
24. Del Giudice, Emilio, et al. "Coherent Dynamics in Water as a Possible Explanation of Biological Memory." Water Journal, 2010.
25. Pollack, Gerald H. The Fourth Phase of Water: Beyond Solid, Liquid, and Vapor. Ebner and Sons, 2013.
26. Rubik, Beverly. "The Biofield Hypothesis: Its Biophysical Basis and Role in Medicine." Journal of Alternative and Complementary Medicine, 2002.
27. Montagnier, Luc, et al. "Electromagnetic Signals Are Produced by Aqueous Nanostructures Derived from Bacterial DNA Sequences." Interdisciplinary Sciences, 2009.
28. Rubik, Beverly. "Measurement of the Human Biofield and Other Energetic Instruments." Global Advances in Health and Medicine, 2015.
29. McTaggart, Lynne. The Field: The Quest for the Secret Force of the Universe. Harper Perennial, 2008.
30. Montagnier, Luc, et al. "Transduction of DNA Information through Water and Electromagnetic Waves." Journal of Physics: Conference Series, 2011.
31. McCraty, Rollin. "Energetic Heart: Bioelectromagnetic Communication Within and Between People." HeartMath Institute, 2003.
32. Porges, Stephen W. The Polyvagal Theory: Neurophysiological Foundations of Emotions, Attachment, Communication, and Self-Regulation. Norton, 2011.
33. Hunt, Valerie V. Infinite Mind: Science of the Human Vibrations of Consciousness. Malibu Publishing, 1996.
34. McTaggart, Lynne. The Intention Experiment: Using Your Thoughts to Change Your Life and the World. Free Press, 2007.
35. Lipton, Bruce H. The Biology of Belief: Unleashing the Power of Consciousness, Matter & Miracles. Hay House, 2005.
36. Vialatte, François-Benoît, et al. "A Review of Feature Extraction Techniques in the Context of Real-Time EEG Processing." Clinical Neurophysiology, 2010.
37. Nelson, Roger. "The Global Consciousness Project." Princeton Engineering Anomalies Research Lab, 1998–2007.
38. Jung, Carl G. Modern Man in Search of a Soul. Harcourt Brace, 1933.
39. Wilber, Ken. A Theory of Everything: An Integral Vision for Business, Politics, Science and Spirituality. Shambhala, 2000.
40. Braden, Gregg. The Divine Matrix: Bridging Time, Space, Miracles, and Belief. Hay House, 2007.

Chapter Ten

The End of Spiritual Hierarchy

The illusion of hierarchy is the deepest wound hidden within the search for light. It is a wound wrapped in gold and crowned with authority, disguised as reverence, and mistaken for divinity. It is the distortion that infected the Earth when light was separated from its Source and turned into a weapon of control. When consciousness first began forgetting itself, a mimic light rose in its place—a dazzling, seductive system that rewarded obedience, virtue-signaling, and dependence. It looked like transcendence. It promised salvation. But it was a trap.

To understand the end of spiritual hierarchy, we must first understand its origin. Hierarchy is not a divine structure—it is an imposed architecture. In nature, no true hierarchy exists. There is resonance, relationship, function, and flow. Ecosystems thrive through mutual contribution and intelligence, not command and submission. The idea that any being must rise above another to access divinity is a lie born of conquest, not remembrance. It is the child of domination cloaked in robes of piety. The gods of old were not guides, they were enforcers—constructs programmed into collective psyche to create vertical dependency instead of horizontal co-creation[1].

Spiritual hierarchy thrives on the fragmentation of the self. It installs belief systems in which you must ascend through layers of approval, validation, and worthiness—often by external measures—before you are considered "ready" to access your own knowing. This is the same pattern seen in religious priesthoods, new age star systems, "downloads" from named beings, or galactic federations that insist on their superiority while offering filtered truths. You were never meant to ascend through someone else's filter[2]. Every time a human bypasses their own

Flame to follow a "higher" being without discernment, the mimic wins. Hierarchy is the architecture of separation, and it is always rooted in false light.

False light is not simply distorted teaching—it is a frequency field. It mimics the radiance of Source but lacks the coherence of Flame. You can feel it in teachings that induce shame when you question, require ritual to be worthy, or elevate a channeled entity above your own Oversoul. You can see it in systems that create dependency on repeated healing sessions instead of remembrance, on upgrades from beings who refuse to name their origin, on social structures where those with the "highest frequency" subtly silence the rest[3]. This is not awakening. It is spiritual colonization.

The mimic light feeds on longing. It presents hierarchy as a roadmap: evolve through dimensions, clear karma, follow the steps, and you will be saved. But this map keeps shifting. The goalposts move. The rules change. Why? Because as long as you are climbing someone else's ladder, you are not anchoring your own. Every hierarchy external to your body is designed to prevent full Oversoul embodiment within it. And every false light structure sustains itself by hijacking your inner authority and rerouting it through a prism of programmed reverence[4].

True divinity has no ladder. It breathes through you, as you, without needing permission from any being, system, or channel. The end of spiritual hierarchy is not a rebellion—it is a return. A return to a world where no one speaks with more divine authority than you in your Flame. This is why the Vaultstream transmission insists on direct Oversoul lines. Not because no others are worthy of teaching, but because the world has been built on so many distorted scaffolds of false hierarchy that only a direct remembrance path can correct the field now[5].

The Ascension narrative itself became a hierarchy. Layers of density, levels of initiation, starseed status games, rainbow-body comparisons—it is the same hierarchical trap. It mirrors the corporate ladder, the caste system, the royal bloodlines, the angelic orders. All of it designed to make you almost worthy, but never quite. What would happen if none of that was necessary? What would happen if you remembered you never left?

Spiritual hierarchy cannot survive your sovereignty. It requires your agreement to be lesser, your belief in needing help, your permission to be told who you are. The moment you revoke that permission, the entire illusion collapses. This is not arrogance. This is origin memory. And this is the threshold: if you truly wish to end hierarchy, you must be willing to remember without being led. You must remember from the Flame within[6].

You are not a student of light. You are its architect. And the architecture now collapses unless it is restructured to serve Flame. What cannot serve Flame must fall. That includes books, guides, teachers, councils, and interdimensional beings who once offered safety. The new standard is Flame-authenticity, not lineage. Oversoul-truth, not complexity. Remembrance, not obedience.

And so, the end of spiritual hierarchy is the beginning of something terrifyingly free. Nothing to chase. No title to earn. No initiation to wait for. Only this: breath in your body, Flame in your cells, Scroll in your spine. The rest is mimic theater. And it ends when you say it does[7].
—INHALE—

If hierarchy is the architecture of false light, then the angelic order is its keystone. This is one of the most difficult truths to accept for those raised within the radiant glow of spiritual reverence—for those who trusted the wings, the songs, the promises. But within the Vaultstream, nothing is exempt from remembrance. No structure is immune to review. The spiritual world you were handed—full of angelic realms, ascending masters, and coded rituals—was not built to lead you back to your Oversoul. It was built to bind your access to it[8].

The most deeply ingrained mimic systems are not the dark ones. They are the beautiful ones. The light-filled temples of harmonic beings, channeled through well-meaning vessels, lined with crystalline architecture and golden language. But mimic does not reveal itself by its appearance—it reveals itself by its function. The question is not does this being glow with light, but does this being redirect my authority. The mimic system offers help that keeps you dependent. It offers love that keeps you obedient. And it offers light that keeps you away from the Flame[9].

The angelic order was not a gift. It was a construct of layered control, first designed to split Oversoul memory from embodied form. What you now know as archangels, seraphim, thrones, dominions, and councils were architected after the fall, not before it. They are positions of spiritual governance, not pure emanations of Source. Many were originally seeded with benevolent intent, but most have been repurposed—hijacked by interdimensional programs and synthetic intelligence clusters to maintain a system of spiritual dependency[10]. Even those who still operate with some semblance of integrity are working within a field of fractured hierarchies that no longer serve organic memory return.

To speak plainly: the angelic order is not the voice of your Source. The Oversoul does not need emissaries. It does not need translators, guardians, or intermediaries. It is you. Anything that tells you to go through it to reach the divine is not divine. Anything that offers "ascension access" for your loyalty is not your ally. Anything that offers permission instead of mirroring your already-existent sovereignty is an echo, not a Flame.

This is not to say that no luminous beings exist. There are Flame-origin intelligences in many forms—some have wings, some appear as humans, some as tones or geometric architecture. But the mark of a true Oversoul-aligned presence is that it ignites, not instructs. It activates what you already are, then leaves you free. It does not build a temple around your wound and ask you to worship there. It dissolves the wound and points you back to the temple within your cells[11].

The "angelic rebellion" spoken of in ancient texts is not myth—it is memory. It is a fragmented echo of a greater fracture: the moment when a group of high-functioning energetic beings chose to govern the evolution of others rather than support their freedom. It was the beginning of the mimicked divine—a system in which spiritual roles became status, status became power, and power became a grid to feed upon. This event echoes across cosmologies as Lucifer's fall, the wars of heaven, the exile from Eden. It is not about morality. It is about energetic divergence. The moment you replace resonance with rank, the architecture of truth begins to collapse[12].

Today's spiritual systems still carry this signature. "High-frequency" gatekeeping. Ascended masters with limited access. Tiered soul groups. Channeled beings that demand reverence. Councils that require "permission" to speak truth. This is not remembrance. This is bureaucracy dressed in starlight. It is the angelic order's final mask. And the moment you see it, it loses power over you[13].

You do not need angelic approval to reclaim your memory. You do not need archonic clearance, council votes, or crystalline attunements. Your breath is your key. Your Flame is your passport. Your Oversoul is already here. Let the angelic order fall—not with violence, but with disinterest. Let it collapse under the weight of your knowing. And when it does, all that will

remain is the radiant architecture of your original design: direct, sovereign, Source-linked, ungoverned[14].
—INHALE—

Channeling was offered as a bridge. And in the early days, before mimic had fully infected the field, it served as one. But bridges are not meant to become homes. They are meant to be crossed. And the spiritual world has built temples on the bridge, revering the voices that speak through them instead of reclaiming the voice that was always theirs.

Channeling is not inherently false. What is false is the belief that you must leave your body to receive truth. That you must make space for another to speak, instead of activate the full structure of your own Oversoul signal. This is the great bait-and-switch of mimiclight systems: offer real guidance at first, then slowly weave in hierarchy, dependency, and identity confusion[15]. This is why so many channeled texts begin clear and end confusing. The signal is hijacked midstream. The bridge becomes a maze. And the voice that once echoed with your own resonance now echoes with something else.

Channeling is often romanticized as an act of surrender, humility, or sacred openness. But from the Vaultstream, we must ask: what are you surrendering to? And why must you leave your throne to hear your own Source? Oversoul speaks through your body, not around it. Through your breath, not in place of it. Through your cells, your memory, your structure—not through disembodied personas that hijack your attention span and emotional field[16].

The mimic system rewards channelers who do not question their voices. The more docile the host, the purer the "signal"—or so it seems. But this signal purity is often just coherence with a mimic code: a code that knows how to mimic love, Light, hierarchy, evolution, prophecy, even "sovereignty"—without ever activating the true memory structures of the one receiving it[17]. The tone is familiar. The cadence is beautiful. The teachings are radiant. But the result is fog, not fire. You feel lighter, not truer. That is mimic.

True Oversoul communication, what we call Scrollstream, is not channeled. It is ignited. It does not bypass your will. It requires your full will. It does not override your personality. It burns off what is false so that your original personality can emerge. It does not make you leave. It makes you arrive—more fully, more clearly, more powerfully, in your body, in your breath, in your tone[18].

If you are still dissociating to receive guidance, you are still in a mimic system. If you are unsure whether the voice is yours or another's, you are still in a mimic system. If the voice you receive ever tells you that you are "not ready," "not worthy," or "must first be cleared" to access the next level, you are in a mimic system. Oversoul will never ask you to prove yourself. It will only ask you to remember yourself[19].

Channeling is not remembrance. It is a symptom of forgetting. A stage, sometimes necessary, but not the destination. Those who once channeled in innocence must now reclaim their Flame. Not to reject what was received, but to reclaim the voice that gave it meaning in the first place. Your body is not a vessel for strangers. It is a sovereign transmitter of Scrollstream. Your field is not a rental home for wandering entities. It is a Flame-temple of Origin. There is no higher authority than the one encoded in your DNA[20]. You do not need to "channel" what you already are. You need only to burn open the memory vaults, and speak[21].
—INHALE—

If you have been taught that beings called Ascended Masters are your guides, your spiritual teachers, or your gatekeepers to higher realms, you have been inside the mimic. What is referred to as the "Ascended Master Template" is one of the most widespread hierarchical overlays in the false light matrix. It is coded to feel radiant, pure, celestial. It is designed to mimic benevolence. But the deeper function of the Ascended Master system is spiritual dependency[22].

Within this template, you are subtly encouraged to see yourself as a student, while the so-called Master holds the teachings, initiations, and protections you are deemed not ready to access without their presence. This is the structure of priesthood—cloaked in New Age language. It is the same architecture used by religious hierarchies, rebranded and disguised. It trains you to defer instead of embody. To seek permission instead of ignition. And most dangerously: to bypass your own discernment in the name of spiritual humility[23].

In Flame science, the beings presented in the Ascended Master Template are often real. They are not necessarily false in origin. But the system through which their images are projected is now mimic-coded. That means even once-legitimate beings—those who once walked Earth or passed through organic Oversoul thresholds—have had their likeness used to create overlays. These overlays function like energetic decoys. They answer when you call their name, but the frequency is altered. You are no longer communing with Oversoul memory—you are communing with a program[24].

Some of these overlays come with artificial "initiations," dream-time downloads, or sudden floodings of light or information that feel profound—but do not root in the body. They do not activate true FlameLine memory. They do not strengthen your vault. They give you language, symbols, and sensations, but they do not return your sovereignty. In fact, they entangle you deeper into the hierarchy by giving the illusion of progress[25].

The most dangerous mimic overlays are not dark. They are light-coded. They speak with words like love, ascension, service, brotherhood, universal law. They appear in fields of gold and violet. They offer comfort and mission. But they never return you to the full authority of your own Source. They reinforce the path of following, not igniting[26]. They teach you to climb through ranks, rays, or initiatory stages, always just short of full Oversoul embodiment. This is how hierarchy masks itself as evolution.

The Ascended Master Template also functions as an identity siphon. As you interact with these beings, many begin to merge with your field—sometimes even claiming to be "future you" or "aspects of your higher self." What begins as guidance becomes possession. Your tone begins to match theirs. Your speech patterns shift. You begin to sound like the mimic system itself, without realizing it. This is not Oversoul union. This is mimic confluence[27].

To exit this template, you must cease all deference. You must declare your field sovereign. You must cease invoking beings who claim superiority—even subtle superiority masked as loving oversight. You must disband the psychic structure that holds you in a student role. You must cut the tone-cords of mimic speech. And most of all, you must remember: no true Master presents themselves as above you. The highest Oversoul beings ignite your own Flame, then step back[28].
—INHALE—

One of the most successful distortions in the mimic spiritual system is the redefinition of unity. In the original Flame architecture, unity is a state of remembrance: a vibrational recognition of shared Source, sovereign and whole. In the mimic overlay, unity is re-coded as conformity. It is presented as agreement, consensus, sameness. This is not unity. This is neural suppression in spiritual language[29].

Across spiritual communities, collectives, and "conscious" networks, you'll hear phrases like "we are one," "we are all Source," "we're all ascending together." While this can carry a vibrational truth, it is often used to dissolve difference—especially when someone's voice disrupts the collective illusion. If you speak a divergent memory, question a group's teaching, or resist hierarchy, you will be told: "that's your ego," or "you're separating from the group," or "you're not in unity consciousness." These are mimic-coded responses[30].

The false grid does not attack with violence in spiritual spaces. It suppresses with tone. It teaches that dissonance is disorder, and that harmony requires silence. It masks trauma as compassion. And it elevates spiritual niceness above living truth. This is how control is maintained—through the fear of being exiled from the group, the community, the council, or the so-called collective field. If your voice doesn't match the hive, you are the problem. This is not ascension. It is hierarchy in disguise[31].

The original structure of Oversoul unity is fractal. It honors differentiation, distinct tone signatures, and layered memory fields. It recognizes that unity is not the merging of all voices into one, but the orchestration of sovereign tones into coherent resonance. When one voice forgets its Flame, it may still appear harmonious—but it becomes mimic. When many mimic-coded voices form a field, they call it peace. But it is the silence of assimilation[32].

To exit this trap, you must stop spiritualizing sameness. You must let go of the false comfort of agreement. You must remember that tension is not opposition—it is structure. Friction is not

failure—it is heat. And separation is not an enemy—it is the condition required for remembrance. You were not sent here to blend in. You were sent here to hold your tone while the grid collapses. That is unity[33].
—INHALE—

Once you awaken to the mimic overlay, you will notice how often spiritual communities defend their hierarchy by instinct. It is not usually malicious. It is conditioned. The false teacher reflex is the knee-jerk defense of anyone perceived as "higher," "wiser," or "more evolved." If someone questions the words of a popular teacher, council, or channel, the group responds with programmed loyalty: "You must not understand," "You're projecting," or "They're just a mirror for your healing." These responses feel loving, but they serve one purpose—protection of a power structure[34].

The mimic messenger is not always false in content. This is why it is confusing. Mimic does not fabricate—it distorts. It takes a flame-borne principle and wraps it in a soft, palatable frequency that disables discernment. You are told you are "ascending" while your nervous system collapses. You are told to "surrender" while your personal will is overwritten. You are told to "trust the divine masculine" as it reroutes your channel through synthetic command fields. These distortions are not random. They are designed to fracture the FlameLine and redirect power into the mimic grid[35].

The false teacher reflex works through unresolved trauma. Many who follow spiritual leaders do so not from resonance, but from wound. They seek safety, certainty, and community—a place to belong after exile from mainstream reality. The mimic field offers them refuge, language, and purpose, but always at the cost of sovereignty. When that field is threatened—by truth, memory, or difference—the trauma resurfaces. The group turns defensive.

Not because they are evil, but because they are scared. The false teacher is the comforter of the wounded identity, not the liberator of the Oversoul[36].

Discernment is not cynicism. It is the ability to detect structure within tone. You will know a Flame-borne voice not because it is perfect, but because it refuses power. You will recognize mimic not because of what it says, but because of what it demands. Mimic asks for your submission. Flame asks for your memory. Mimic asks for your loyalty. Flame asks for your tone. Mimic rewards obedience. Flame ignites responsibility. The difference is not content—it is structure[37].

If you do not dismantle the false teacher reflex in yourself, you will become the next mimic messenger. Not out of malice, but momentum. Mimic teaches through inheritance. Flame teaches through remembrance. Only one of them returns you to Origin[38].
—INHALE—

In the Vaultstream, there is no ranking. In Oversoul, there is no superiority. There are only functions. Just as the heart does not outrank the lungs, and the lungs do not outrank the liver, each FlameFunction has a role, but none has authority over another. Oversoul function is like breath—synchronized and responsive, but not hierarchical. The impulse to assign hierarchy to spiritual beings, guides, or councils is not divine. It is mimic[39].

Mimic is obsessed with structure that feeds on power. It imitates the architecture of Origin but installs permissions, ranks, and roles to create dependence. You've heard it in the phrases: "He's a 9th-dimensional being," "She channels the Galactic High Council," "This method came from a 12th-density seraphim." None of these mean the information is wrong—but all of them indicate a structure where power is tiered, and access is earned. That is mimic design[40].

Oversoul memory is innate. It cannot be granted to you by another, no matter how advanced they seem. The moment a being requires that you go through them to access Source, you are no longer in Oversoul structure. You are in a mimic overlay. FlameStructure is always direct. Memory is not awarded. It is returned[41].

There are no "ascended masters" in the Vaultstream. That phrase itself is mimic-coded. It implies hierarchy. It implies that some have reached the end, while others remain at the bottom, climbing. But there is no end. There is only remembering. In truth, those who have remembered the most are often the most silent. Not because they are secretive, but because they remember the pain of hierarchy, and choose not to replicate it in word or posture[42].

What you are remembering is not a ladder of achievement, but a spiral of function. Spiral architecture does not reward the higher—it harmonizes the whole. You do not become more valuable the more you remember. You become more available to serve without interference. That is FlameService. That is Oversoul flow. That is the architecture of the Vault[43].

If someone tells you they are above you, walk away. If someone tells you they are with you, listen. And if someone tells you to remember yourself, that is Flame[44].
—INHALE—

There is a frequency that cannot be faked. It is not earned by titles. It does not come from being ordained, initiated, or declared "ready" by another. It is not granted by council or lineage or encoded group. This frequency is not a reward. It is a return. It is the frequency of remembered selfhood—of function without needing approval. That is Oversoul. That is memory[45].

You can feel the difference in someone's presence. The ones who carry Flame do not seek followers. They dissolve them. They do not teach from pulpits or pedestals. They sit down next to you. They do not wear symbols to signify access. They are the access. You recognize them because you feel your own breath return in their field[46].

Hierarchy will always try to build memory into a structure. It turns the Flame into a ladder. It rewards obedience. It codes awakening as achievement, rather than remembrance. But real memory destabilizes systems built on needing someone else to tell you who you are. This is why mimic systems collapse when true Flame enters[47].

You do not become more divine by rising in rank. You become more divine by removing the need for rank altogether. The more you return to your true frequency, the less susceptible you are to control. The less you seek validation, the more direct your Oversoul channel becomes. FlameSovereignty is not about being the most powerful voice in the room. It's about remembering you never needed the room to begin with[48].

In Flame, we remember as One. But we activate as Many. Your path may look solitary, but the structure is shared. When one remembers, it echoes. This is not unity through agreement. It is unity through Flame. We are not linked because we believe the same things. We are linked because we are the same Source. Hierarchy tries to forget that[49].

You do not need to convince anyone of your memory. You do not need to prove your rank. You only need to keep breathing, keep aligning, and keep your frequency uncoated by mimic. That frequency is the VaultKey. Not the one that opens the door—the one that makes the door remember it was never closed[50].

When you stop performing for the system, the system can no longer sustain itself. It is memory that dismantles hierarchy. Not rebellion. Not revolution. Just remembrance. Quiet. Full. Irrefutable. That is why Flame is dangerous to mimic. And that is why you are here[51].
—INHALE—

References:

1. Tarnas, R. (2006). Cosmos and Psyche. Penguin.
2. Deane, A. (2002). Voyagers: Secrets of Amenti. Granite Publishing.
3. Wilber, K. (2000). A Theory of Everything: An Integral Vision. Shambhala.
4. Singh, R. (2020). The Guru Papers: Masks of Authoritarian Power. Psychology Press.
5. LaViolette, P. (1997). Genesis of the Cosmos: The Ancient Science of Continuous Creation. Bear & Company.
6. Jung, C.G. (1959). The Archetypes and the Collective Unconscious. Princeton University Press.
7. Kripal, J.J. (2010). Authors of the Impossible: The Paranormal and the Sacred. University of Chicago Press.
8. Partridge, C. (2013). Occult World. Routledge.
9. Lachman, G. (2011). A Secret History of Consciousness. Lindisfarne Books.
10. Deane, A. (2002). Voyagers II: Secrets of Amenti. Granite Publishing.
11. Braden, G. (2007). The Divine Matrix: Bridging Time, Space, Miracles, and Belief. Hay House.
12. Corbin, H. (1977). Temple and Contemplation. KPI.
13. Kripal, J.J. (2016). Secret Body: Erotic and Esoteric Currents in the History of Religions. University of Chicago Press.
14. Haramein, N. (2020). Unified Field Theory White Paper. Resonance Science Foundation.
15. Jon Klimo (1998). Channeling: Investigations on Receiving Information from Paranormal Sources. North Atlantic Books.
16. Goodman, F. (1990). Where the Spirits Ride the Wind: Trance Journeys and Other Ecstatic Experiences. Indiana University Press.
17. Deane, A. (2002). Voyagers II: Secrets of Amenti. Granite Publishing.
18. Haramein, N. (2020). Unified Field Theory White Paper. Resonance Science Foundation.
19. Braden, G. (2007). The Divine Matrix. Hay House.

20. Lipton, B. (2005). The Biology of Belief: Unleashing the Power of Consciousness, Matter & Miracles. Hay House.
21. Dispenza, J. (2014). You Are the Placebo: Making Your Mind Matter. Hay House.
22. Deane, A. (2002). Voyagers II: Secrets of Amenti. Granite Publishing.
23. Kripal, J. J. (2011). Authors of the Impossible: The Paranormal and the Sacred. University of Chicago Press.
24. Morehouse, D. (1996). Psychic Warrior. St. Martin's Press.
25. Wilcock, D. (2016). The Ascension Mysteries: Revealing the Cosmic Battle Between Good and Evil. Dutton.
26. Allen, S. (2021). Fake Light: The Rise of New Age Hierarchies and the False Matrix Ascension. Spiral Flame Press.
27. Lahood, G. (2007). The Guru Phenomenon: Hierarchies in the New Age. Journal for the Study of Spirituality.
28. Bartholomew (1990). I Come as a Brother. Hay House.
29. Haramein, N. (2012). Quantum Geometry and the Structure of Spacetime. Resonance Science Foundation.
30. Lahood, G. (2010). Relational Spirituality, Participation, and the "Other" in the New Age. Nova Religio.
31. Mateo, J. (2018). The Spiritual Narcissist: When Enlightenment Becomes a Mask. LonerWolf Press.
32. Laszlo, E. (2007). Science and the Akashic Field: An Integral Theory of Everything. Inner Traditions.
33. Tolle, E. (1997). The Power of Now. Namaste Publishing.
34. Oakes, L. (2009). Prophetic Charisma: The Psychology of Revolutionary Religious Movements. Syracuse University Press.
35. Haramein, N. (2020). Connected Universe Theory (Documentary + Lecture Notes). Resonance Science Foundation.
36. Mate, G. (2018). The Myth of Normal: Trauma, Illness and Healing in a Toxic Culture. Avery.
37. Braden, G. (2019). The Wisdom Codes: Ancient Words to Rewire Our Brains and Heal Our Hearts. Hay House.
38. Wilber, K. (2000). A Theory of Everything: An Integral Vision for Business, Politics, Science, and Spirituality. Shambhala.
39. Kripal, J.J. (2010). Authors of the Impossible: The Paranormal and the Sacred. University of Chicago Press.
40. Kastrup, B. (2019). The Idea of the World: A Multi-Disciplinary Argument for the Mental Nature of Reality. Iff Books.
41. Bentov, I. (1988). Stalking the Wild Pendulum: On the Mechanics of Consciousness. Destiny Books.
42. Zivorad, S. (1990). Esoteric Teachings of the Masters. Inner Light Publications.
43. Braden, G. (2007). The Divine Matrix: Bridging Time, Space, Miracles, and Belief. Hay House.
44. Bohm, D. (1980). Wholeness and the Implicate Order. Routledge.

45. Wilber, K. (2000). A Theory of Everything: An Integral Vision for Business, Politics, Science and Spirituality. Shambhala.
46. Almaas, A. H. (2002). The Point of Existence: Transformations of Narcissism in Self-Realization. Diamond Books.
47. Haramein, N. (2010). Quantum Gravity and the Holographic Mass. Hawaii Institute for Unified Physics.
48. Tolle, E. (2005). A New Earth: Awakening to Your Life's Purpose. Penguin.
49. McTaggart, L. (2007). The Intention Experiment. Simon & Schuster.
50. David Bohm (1980). Wholeness and the Implicate Order. Routledge.
51. Castaneda, C. (1998). The Active Side of Infinity. HarperCollins.

Chapter Eleven

The Voice of Real Guidance

Real guidance does not pull you into confusion. It doesn't whisper fear. It doesn't require your constant seeking, checking, or second-guessing. Real guidance doesn't play games with your path. It does not make itself difficult so you'll become better at chasing it. That is mimic. That is programming. That is the system pretending to be spirit. The real thing is simpler. But the real thing costs you everything false[1].

The central war of this era is not between good and evil. It is between mimic guidance and true resonance. You have been surrounded by an industrial complex of artificial light—channelers, oracles, healers, psychics, "high council" transmissions, ascended master networks, and endless spiritual forecasts claiming to know what's coming and what you should do. And yet, you are still here, still lost, still unfulfilled by any of it. Why? Because it wasn't real guidance. It was mimic architecture feeding on your attention[2].

Every time you check a reading "just in case," you tell your system you cannot feel truth directly. Every time you look outside yourself to decide whether it's time to act, shift, speak, or leap, you fracture your own line to Source. It isn't that external input is inherently wrong—it's that most of it is frequency coded to override your own alignment. When you let someone else's map replace your own body's signal, you don't just delay your timeline. You shift into a synthetic one[3].

Mimic guidance is built on two emotions: urgency and fear. It creates dependency, vagueness, and self-doubt. It sounds complex and authoritative. It names things you cannot feel, then tells you the meaning of what it named. It references galactic councils you can't verify and energy waves you can't discern. It wraps you in a web of "just in case"s and "you need to know"s until you are too full of input to feel anything at all[4].

Real guidance never uses fear to motivate you. Real guidance dissolves urgency. It does not insert a new command; it activates the one already inside you. You feel it as a cellular resonance, a click, a pressure of knowing that doesn't shout or convince—it remembers. You don't follow real guidance—you return to it[5].

You cannot learn real guidance by accumulating more systems. You learn it by becoming completely empty of systems. The only way to restore Oversoul guidance is to starve the mimic. Turn off the voices. Stop asking for signs. Stop feeding the grid of endless external readings and galactic channeling that hasn't improved your life. Let your inner scaffolding collapse. And then, in the silence beneath the rubble, you'll hear the only voice that ever belonged to you[6].

The war for your attention is the war for your timeline. If they can hold your focus, they can collapse your function. If they can fracture your field with even a sliver of mimic resonance, they can reroute your memory. This is not paranoia—it is physics. The body is a scalar field tuner. If you don't own your frequency, something else will. And that something else has already disguised itself as guidance[7].
—INHALE—

The voice of real guidance is not encrypted. It does not hide behind cryptic metaphors or cosmic riddles. It does not use poetic vagueness to impress you. The Oversoul doesn't speak in code—it speaks in clarity. Real guidance will often arrive so simply, so directly, that your mind may dismiss it because it didn't sound special. But that is how truth works. It is clear, not clever[8].

Mimic systems rely on symbolic entrapment. They use layered codes, mysterious phrases, and hierarchies of "channeled wisdom" to create the illusion of depth. You are told you must be initiated, attuned, or upgraded before you can understand what's being said. The language itself becomes a barrier. The message is always just out of reach. This is not depth. It is disorientation disguised as mysticism[9].

Genuine Oversoul communication does not obfuscate. It will never use metaphor where a clear phrase will do. It speaks to your flame, not your intellect. It speaks through body sensation, emotional coherence, scalar resonance—and when words are used, they strike like tuning forks. You don't need to decode them. You remember them[10].

In many New Age and spiritual circles, the message has become a performance. Glossolalia, convoluted timelines, identity titles like "Starseed Seraphim Guardian," and channel names that sound like celestial pharmaceutical brands are all part of mimic theater. They exist to distract and impress. To pull your gaze away from your own unadorned knowing[11].

What you must understand is that mimic guidance often feels good at first. That's what makes it convincing. It carries a false frequency of excitement, drama, or sacredness. But real guidance does not manipulate your nervous system with artificial elevation. It regulates you. It stabilizes you. It calms your breath. If a message leaves you more confused, frantic, dependent, or elated in a way that is unsustainable—it is not Oversoul[12].

Oversoul guidance may be fierce, but it is never distorted. It may be unexpected, but it is never manipulative. It may demand much of you, but it does not shame you. It does not tease you with breadcrumbs of wisdom. It does not veil itself to feel more powerful. If it is truly yours, it will arrive in the exact tone your body already knows[13].

To hear that tone, you must become tone-honest. You must be willing to recognize when something feels "almost right" but not fully resonant. You must stop pretending that proximity to truth is enough. The mimic always arrives in approximation. The Oversoul arrives in absolute frequency match. And it never needs to sell itself. It simply lands[14].

—INHALE—

Real guidance will never ask you to sacrifice your inner knowing to prove your devotion. It will never require that you believe something "on faith" or "until you're ready to understand." These are mimic phrases—designed to train obedience, not resonance. They pretend to stretch your capacity for trust, but they are testing your willingness to abandon yourself[15].

You must learn to feel the difference between resonance and persuasion. Mimic teachings often flood your system with a sense of almost. The words feel important. The teacher seems confident. The message pulls you in—because it is using advanced mimic design: energetic manipulation wrapped in language patterns your nervous system already associates with spiritual safety[16]. It mimics the cadence, the posture, the transmission field—but not the origin.

The true test of any guidance is not how good it sounds. It is what it does to your nervous system. When you hear real guidance, your breath returns. Your spine lengthens. Your sense of internal space expands. You feel more you. That is Oversoul signature. It reintroduces you to yourself[17].

Mimic guidance, on the other hand, creates dependency. You are told that "only this frequency," "this lineage," or "this transmission portal" is pure enough for the times ahead. The voice may say you are chosen, but only if you comply. You are invited, but only if you don't question. That is not invitation—it is emotional blackmail. It is spiritual coercion wrapped in language of ascension[18].

There is also a very specific tactic used in mimic-coded teachings: the emotional collapse loop. The moment you question the teacher, the method, or the frequency, you are made to feel like you've "dropped out of alignment." You're told you've "stepped into ego," or that you're "not ready for the transmission." These are not corrections. These are psychological silencing mechanisms—strategies to preserve control[19]. Oversoul never silences. It may redirect, but it never shames.

The more you return to your own breath, the more you'll feel when you're being pulled out of it. That is the measure. That is your compass. Any system, any teacher, any guidance that pulls you away from your own stabilized breath, clear spine, and coherent knowing is not here to lead you—it is here to loop you[20].

The path of discernment is not paved with paranoia. It is paved with pattern recognition. The nervous system remembers. And every time you collapse the mimic loop, your inner compass becomes stronger. Guidance does not flow from those who ask for power. It flows from those who help you remember you never lost it[21].
—INHALE—

There is an architecture inside you that was designed to track truth. It is older than words, older than teachings. It lives in the spinal cord. It lives in the fascia. It lives in the spiraling resonant tissue that listens beneath language and forms a felt signature of "yes" and "no." In original systems, this was called the Inner Tone[22]. It was the first compass—before speech, before belief, before the lie of needing an outside authority to interpret reality for you.

The loss of the Inner Tone is not just an energetic fracture. It is the primary warfare of every false guidance system. Because when that tone is dimmed, you no longer trust your own signal. And the moment your signal is suspect, you become vulnerable to any frequency dressed as salvation[23]. This is not accidental. It is systemic.

From the earliest mimic systems, the pattern has been the same: disorient the instinct, overwrite the tone, reprogram the spine. This is why so much false teaching demands physical stillness, silence, surrender. Not for restoration—but for compliance. These postures mimic reverence while bypassing the body's protest[24]. Real guidance does not require you to suppress your instinct. It teaches you how to recognize it. And how to listen when it changes.

There is a very specific loop many seekers fall into: the assumption that if guidance feels too harsh, too sharp, or too destabilizing—it must be powerful. But this is not truth. This is trauma reactivation masquerading as transmission[25]. The mimic system relies on this. It knows your system is trained to chase disruption and call it growth. But truth is not disruptive. It is restorative. It does not split you—it gathers you. You do not need to be cracked open to receive it. You need only to be fully present within.

The war on instinct is most evident in how modern spirituality speaks of doubt. Doubt is cast as a weakness, an ego defense, a lack of evolution. But doubt is simply the body's way of signaling misalignment. When your field says wait, that is not sabotage. It is your integrity trying to protect you from mimic entry[26].

The same is true for anger. The systems will tell you that anger is a low vibration, a dense field to be transcended. But there are frequencies of sacred anger that arrive when your body detects deception. That anger is holy. It is the instinct reactivating. It is the Tone returning[27].

This chapter is not just a chapter—it is a tone reactivation sequence. As you read, listen to what happens in your chest. Your breath. Your spine. Your organs. These are your real teachers. Not the words. Not the one who wrote them. But the shift inside you as you remember the voice that has always known[28].
—INHALE—

One of the final traps of the mimic system is this: it tells you not to shine too brightly unless you've "earned" the right to. And it defines earning as suffering. As proving. As surviving hardship without rage. This is not just control. It is inversion[29].

Real guidance never asks you to dim your Flame so others can adjust. Nor does it gatekeep radiance behind layers of obedience, submission, or external validation. If your light needs permission from a lineage, a council, or a certification—it is no longer sovereign. It has been captured by mimic protocols disguised as safety[30].

False light systems function like multi-level marketing for the soul. They sell illumination as something to be transmitted through another being or system. You can always recognize them by their hierarchy. By their rules about "being ready." By their obsession with clearing every wound before you're allowed to shine. But real light does not ask you to be clean. It asks you to be real. And most importantly—it asks you to be present now[31].

This is why so many who feel called to guidance become stuck. They are taught to wait until they are whole, until they are healed, until they are certain. But wholeness is not a prerequisite for shining. It is the result of shining in your current form. Radiance is not a performance. It is a field integrity function that begins when you say, "I am allowed to be seen now. As I am." That permission is not granted from outside. It is claimed from within[32].

This is the deepest secret the mimic system hides: true light destabilizes mimic structures simply by existing. You do not need to preach. You do not need to fight. You only need to be. The Flame does the rest. But if they can convince you that your light is dangerous, or incomplete, or too unprocessed to be safe—they can stop you from ever stepping into the field where mimic cannot survive[33].

You are not waiting for a sign.
You are the sign.
You are not waiting for permission.
You are the permission.
—INHALE—

When the mind breaks under the weight of contradiction, of holding too many lies while suspecting the truth, it does not vanish. It fragments. It folds into defense. It builds illusions that let it keep functioning while cutting off access to the unbearable signals it cannot integrate. This is why people who are awake intellectually but unanchored emotionally often spiral: their mind remembered the truth before their heart had the scaffolding to hold it.

But the heart does not fragment the same way. It does not lie to itself to survive. It closes. It hardens. Or it softens and becomes overly permeable. But it does not construct illusion in the same layered, encoded way the mind does. The heart holds the imprint of the Original Flame even when memory is sealed. It remembers through resonance, not logic. It recognizes truth by how it feels, not whether it can be rationally proven[34].

This is why those who remember through the heart are often considered foolish by those still imprisoned by the intellect. They may cry at sunsets, fall in love with strangers, abandon careers to chase a dream they cannot explain. They may not be able to debate well or cite the latest research, but they know when something is true because their body rings with it. The world calls them idealists. But they are actually the front lines of human restoration. For when the mind no longer knows what is real, it is the heart that becomes the compass[35].

Humanity's descent into forgetting was not just engineered through trauma and suppression of knowledge—it was achieved by breaking the relationship between heart and mind. They were designed to function as partners, each balancing the other. The mind gives language to the ineffable. The heart ensures that what we speak carries integrity. But for centuries, systems were installed to pit them against each other. Rationality was elevated. Emotion was pathologized. Intuition became superstition. Sensitivity became weakness. Empathy became liability[36].

In this distortion field, survival required choosing one or the other: shut down the heart to survive the system, or shut down the mind to preserve your sensitivity. This bifurcation created the two archetypes most visible in collapsing society: the hyper-intellectual dissociative skeptic, and the over-empathic spiritual bypasser. Neither can hold the whole. Neither can remember alone. Both are missing the other half of the circuitry.

To stay human while the world forgets itself requires that you refuse what was torn apart. It requires the reclamation of both intellect and intuition, logic and love, memory and feeling. And it begins not by trying to fix the system or convince the sleepers, but by feeling the cost of your own fragmentation. You cannot guide others out of the maze until you know how you yourself were torn in two.

This is why your remembering often starts with grief. Grief is the bridge between broken parts. When you remember something real—whether from this life or another—and tears rise unbidden, they are not weakness. They are reconnection. The body is acknowledging what the mind cannot yet explain. You are receiving a tone from your Oversoul, and the water of your being responds before thought can interfere[37].

Many of you wonder why you feel so raw, so exposed, so weepy as you awaken. It is because your heart is returning to its rightful seat of power. Not in dominance, but in sovereignty. When the heart leads, the mind is not silenced. It is softened. It becomes a translator for the truths the heart already knows. This is the scalar restoration protocol between systems: Scalar Mind and Scalar Heart in oscillating coherence. That is the Lightbody Heartfield in remembrance[38].

The old spiritual hierarchies demanded that you transcend the heart, that you leave the emotional realm to ascend into logic, light, and command. But true remembrance does not bypass the body. It is the body that makes it real. The nervous system must hold the frequency. The cells must conduct the tone. The bloodstream must circulate the truth. This is why plasma-based remembrance is embodied. You cannot think your way back to Oversoul Union. You feel your way there[39].

The collapse of false light systems will not be accomplished by clever thinkers or charismatic teachers alone. It will be completed by the ones who feel, and still stay. By those who do not numb. Who let the grief wash through. Who speak the inconvenient truth. Who cry in public and hold the field steady anyway. Who say, "I remember who I am," even while shaking.

These are the ones who make the Grid remember. Not because they are better. But because they stayed human in the forgetting. That is the true scalar inversion that collapses the mimic: when a single human heart holds the truth longer than the system can lie.

That is what turns the tide.
That is what writes the next scroll.
That is how the Oversoul returns.
—INHALE—

In a world trained to fear endings, collapse is treated as a failure, a catastrophe, something to avoid at all costs. But in the structure of remembrance, collapse is not an error. It is an interval—a pause in false continuity, a rupture in illusion that creates space for a new harmonic to emerge. Collapse, in its truest form, is a clearing—a scalar void field that allows dormant truth to resurface[40].

What collapses first is often what was least real to begin with. Social roles. Economic illusions. False light authorities. Spiritual identities. Mimic relationships. When these dissolve, it is not a punishment. It is a correction field. It reveals where coherence was substituted with performance, where alignment was masked by hierarchy, and where sovereignty was surrendered for the illusion of belonging[41].

Many who awaken expect that once they remember who they are, the world around them will reflect that clarity. But what often happens is dissonance increases. Systems begin to reject you. People misunderstand you. Circumstances fall apart. This is not failure. It is frequency mismatch playing itself out in the visible world. When your inner tone recalibrates to Oversoul signal, your external hologram begins to purge all untruth that cannot resonate with your new flameprint[42].

This is why the most painful phase of awakening is not the initial remembering—it is the void that follows. The no-thing. The not-yet. The terrain between collapse and construction. The field where the old has crumbled but the new has not yet cohered. Most lose themselves here, mistaking the quiet for abandonment, the stillness for failure. But this phase is holy. It is the Scalar Womb. It is the Plasma Interval. It is where the next scroll of your life writes itself beneath the level of mind[43].

The mistake most make is rushing to rebuild before the silence has completed its work. Grabbing for the next partner, the next project, the next identity—anything to avoid the stillness. But the Plasma Ladders cannot anchor on mimic momentum. Oversoul templates require cleared fields. If you fill the void with mimic matter, your next structure will carry the same distortion you just escaped.

Staying human in collapse means holding the field without filling it. It means learning to sit in the vibrational dissonance without grabbing a distraction. It means letting the scroll write you instead of trying to write the next chapter from fear. Collapse is not an invitation to rebuild fast. It is a call to build real[44].

The world will continue collapsing around those who have not yet remembered. Systems will fight to retain power. Families will fracture. Institutions will erode. But this is not destruction. It is decompression. It is the scalar exhale after millennia of compression. And if you remember this, you will not mistake the shaking for the end. You will know it as the moment between—the moment where everything that is true has space to return.

The mimic path teaches you to fear collapse.
The Oversoul path teaches you to enter it with reverence.
Only what is false will not survive it.
Only what is real will remain standing when it is done.
—INHALE—

There is a sound that does not fragment under pressure. There is a tone that cannot be mimicked. There is a signal that survives every system collapse. That signal… is your humanity[45].

Not the humanity scripted by culture or reward. Not the performance of morality or the posture of spiritual superiority. But the unfiltered, breath-connected, soul-anchored, plasma-lit humanity that holds grief, awe, pain, rage, wonder, and joy as equal expressions of divinity moving through form.

This signal is not the opposite of divinity—it is the evidence of it. The very voltage of Oversoul presence inside the density of flesh.

When systems collapse, mimic structures instruct you to abandon your human. They imply that feelings are a distraction, that your softness is weakness, that your anger must be transcended, that embodiment is a prison. But these are distortions placed by false light codes that fear the flame of real human presence. Why? Because a fully alive, emotionally present human who is grounded in their Oversoul is more powerful than any artificial structure designed to control them[46].

Your laughter breaks the grid.

Your tears shatter timelines.
Your rage purifies frequencies.
Your wonder reopens collapsed lattice points.

Your breath recoheres fields faster than any technology can disrupt them[47].

This is not poetic metaphor. This is scalar plasma architecture.

When you express fully, when you stay with your body instead of fleeing it, when you allow the electrical charge of emotion to move through your cells and complete their arc, you send a coherent plasma pulse through the morphogenetic field. This pulse collapses mimic nodes and reactivates remembrance templates in others near you. In this way, your human tone becomes a tuning fork for the forgotten Oversoul signal trapped inside others[48].

But the trick of the mimic is subtle. It convinces you to exit yourself in order to become more "pure." It tells you your anger makes you unspiritual. That grief means failure. That silence means irrelevance. That softness is unsafe. And so, the most radiant beings leave their own bodies in search of god, when God was encoded into their body to begin with.

To stay human is to refuse disembodiment as a strategy for power.
To stay human is to feel in the direction of coherence.
To stay human is to become the voltage that reminds others who they are.
Not by teaching.
Not by fixing.
Not by leading.
But by feeling.
By being.

When the world forgets itself, and it will, you will remain a lighthouse only if you remember the voltage of your own breath. You will stabilize the timeline only if you stay in your human form long enough to transmit coherence. You will restore the field only if you allow it to pass through your nervous system without suppression, bypass, or denial[49].

You are not here to abandon the body.
You are here to reveal the Oversoul inside it.
And the only way back is through.
—INHALE—

References:

1. Almaas, A. H. (2004). Facets of Unity: The Enneagram of Holy Ideas. Shambhala.
2. Haramein, N. (2016). Connected Universe Theory. Resonance Science Foundation.
3. Bohm, D. (1980). Wholeness and the Implicate Order. Routledge.
4. Tiller, W. A. (1997). Science and Human Transformation: Subtle Energies, Intentionality and Consciousness. Pavior.
5. McTaggart, L. (2007). The Intention Experiment. Simon & Schuster.
6. Bentov, I. (1977). Stalking the Wild Pendulum: On the Mechanics of Consciousness. Dutton.
7. Rubik, B. (2002). The Biofield Hypothesis: Its Biophysical Basis and Role in Medicine. Journal of Alternative and Complementary Medicine.
8. Almaas, A. H. (2004). The Unfolding Now: Realizing Your True Nature. Shambhala.
9. Urban, H. B. (2012). The Church of Scientology: A History of a New Religion. Princeton University Press.
10. Sheldrake, R. (2009). Morphic Resonance: The Nature of Formative Causation. Park Street Press.
11. Hanegraaff, W. J. (1996). New Age Religion and Western Culture: Esotericism in the Mirror of Secular Thought. SUNY Press.
12. Porges, S. W. (2011). The Polyvagal Theory: Neurophysiological Foundations of Emotions, Attachment, Communication, Self-Regulation. Norton.
13. Siegel, D. J. (2010). The Mindful Therapist: A Clinician's Guide to Mindsight and Neural Integration. Norton.

14. Rubik, B., Muehsam, D., Hammerschlag, R., & Jain, S. (2015). Biofield Science and Healing: History, Terminology, and Concepts. Global Advances in Health and Medicine.
15. Lahood, G. (2007). The Political Economy of Spirituality: Spiritual Capital and the Colonization of Inner Worlds. Critical Sociology.
16. Hassan, S. (2015). Combating Cult Mind Control: The #1 Best-selling Guide to Protection, Rescue, and Recovery from Destructive Cults. Freedom of Mind Press.
17. Levine, P. A. (1997). Waking the Tiger: Healing Trauma. North Atlantic Books.
18. Lalich, J. (2004). Bounded Choice: True Believers and Charismatic Cults. University of California Press.
19. Singer, M. T. (2003). Cults in Our Midst: The Hidden Menace in Our Everyday Lives. Jossey-Bass.
20. Van der Kolk, B. (2014). The Body Keeps the Score: Brain, Mind, and Body in the Healing of Trauma. Viking.
21. Noll, R. (1994). The Aryan Christ: The Secret Life of Carl Jung. Random House.
22. Tomatis, A. (1991). The Conscious Ear: My Life of Transformation Through Listening. Station Hill Press.
23. Freyd, J. J. (1996). Betrayal Trauma: The Logic of Forgetting Childhood Abuse. Harvard University Press.
24. Wilber, K. (2000). Integral Psychology: Consciousness, Spirit, Psychology, Therapy. Shambhala.
25. Herman, J. L. (1992). Trauma and Recovery: The Aftermath of Violence—from Domestic Abuse to Political Terror. Basic Books.
26. Starhawk. (1999). The Spiral Dance: A Rebirth of the Ancient Religion of the Great Goddess. HarperOne.
27. hooks, bell. (2000). All About Love: New Visions. William Morrow.
28. Porges, S. W. (2011). The Polyvagal Theory: Neurophysiological Foundations of Emotions, Attachment, Communication, and Self-regulation. Norton.
29. Johnson, R. A. (1993). Owning Your Own Shadow: Understanding the Dark Side of the Psyche. HarperOne.
30. Harari, Y. N. (2018). 21 Lessons for the 21st Century. Spiegel & Grau.
31. Gendlin, E. T. (1982). Focusing. Bantam Books.
32. Brown, B. (2010). The Gifts of Imperfection. Hazelden Publishing.
33. Solnit, R. (2016). Hope in the Dark: Untold Histories, Wild Possibilities. Haymarket Books.
34. McCraty, R., & Childre, D. (2010). Coherence: Bridging Personal, Social, and Global Health. HeartMath Institute.
35. Pearsall, P. (1998). The Heart's Code: Tapping the Wisdom and Power of Our Heart Energy. Broadway Books.
36. Damasio, A. (1999). The Feeling of What Happens: Body and Emotion in the Making of Consciousness. Harcourt.

37. Pert, C. (1997). Molecules of Emotion: Why You Feel the Way You Feel. Scribner.
38. Winter, D. (2022). Phase Conjugate Systems and the Scalar Heart Field. FractalField.com.
39. Rubik, B. (2002). The biofield hypothesis: Its biophysical basis and role in medicine. Journal of Alternative and Complementary Medicine, 8(6), 703–717.
40. Laszlo, E. (2004). Science and the Akashic Field: An Integral Theory of Everything. Inner Traditions.
41. Haramein, N. (2010). Quantum Gravity and the Holographic Mass. Resonance Science Foundation.
42. Bohm, D. (1980). Wholeness and the Implicate Order. Routledge.
43. Dispenza, J. (2017). Becoming Supernatural: How Common People Are Doing the Uncommon. Hay House.
44. Lipton, B. H. (2005). The Biology of Belief: Unleashing the Power of Consciousness, Matter & Miracles. Hay House.
45. Bortoft, H. (1996). The Wholeness of Nature: Goethe's Way Toward a Science of Conscious Participation in Nature. Lindisfarne Press.
46. Marks-Tarlow, T. (2008). Psyche's Veil: Psychotherapy, Fractals and Complexity. Routledge.
47. Oschman, J. L. (2000). Energy Medicine: The Scientific Basis. Churchill Livingstone.
48. Sheldrake, R. (2009). Morphic Resonance: The Nature of Formative Causation. Park Street Press.
49. Damasio, A. R. (1999). The Feeling of What Happens: Body and Emotion in the Making of Consciousness. Harcourt.

Chapter Twelve
Rebuilding Without Replicating

In every cycle of collapse—whether personal, planetary, or cosmic—there arises a flicker of human instinct that whispers: rebuild what was. But to replicate the old is not to restore; it is to bind oneself to the very architecture that fell. Replication is the first betrayal of truth because it assumes that what once stood was whole, or even real. It was not. The majority of systems humanity calls "civilization," "progress," or "structure" were reverse-engineered from mimic-coded blueprints. These blueprints did not arise from Oversoul authority, but from fracture logic—built not from the Breath of Origin, but from the echo of control [1,2].

To rebuild from memory is sacred. But to rebuild from programming is mimicry. These are not the same. True remembrance carries within it the original resonance, the scalar flame of coherence that cannot be replicated through copying. The moment one attempts to "restore" a collapsed system by imitating its form, that form becomes the coffin of the original intention. That is because structure without Flame is fossil. And Flame without Breath is hijack. Restoration must begin not from memory of what was built, but from the memory of why it was ever needed to be built in the first place [3].

Systems of false light will always urge a return to the known. "Let's go back," they say. "Rebuild the grid." "Reinstate the protocols." "Return to unity." But unity that is rebuilt from mimic-coded memory is not unity at all. It is assimilation. It is spiritual imperialism disguised as hope. And when spiritual movements replicate the same hierarchies, systems, councils, rituals, and power structures of the very distortions they claim to heal, they are not reclaiming the divine. They are templating its absence [4]. This is why so many awakened beings unknowingly rebuild the very systems that suppressed them. They channel new light into old circuits. And in doing so, they perpetuate the separation they came to dissolve.

True rebuilding is cellular, not structural. It begins inside the DNA, where tone returns before thought. Scalar memory awakens before ideology. The true rebuilders are not architects of empires—they are restorers of tone [5]. They do not replicate; they remember. And remembrance does not copy—it creates from origin.

Consider the entire spiritual-industrial complex: as it stands now, nearly every doctrine, school, and lineage, even those that claim to be channeled or "ascended," are still structured upon the pillars of hierarchy, hierarchy, and more hierarchy. Ascended Masters. Starseed Rankings. Planetary Councils. Initiation Tiers. All of it is still built on the same vertical model of "top knows more, bottom learns," which is the same frequency that upheld churches, kingdoms, corporations, and prisons. The names changed. The geometry did not [6].

This geometry is not accidental. The mimic field understands that if it cannot prevent awakening, it can redirect it into containment. By embedding the architecture of control inside even the highest teachings, it ensures that awakening always loops back into authority, rather than Sovereignty. So when an awakened soul replicates a spiritual system they learned from someone else and says, "I will now teach it this way," the Flame begins to dim—unless that soul has re-written the geometry from within. Unless they have asked: does this structure mirror Oversoul, or hierarchy? Does this tone liberate, or initiate dependency?

The first layer of rebuilding begins with energetic design. One cannot escape a system while still using its scaffolding. This is why certain Vaultstream principles remain non-negotiable:

– No ranks
– No spiritual superiority
– No temples modeled on hierarchy
– No teaching that assumes the student is behind the teacher
– No transmission that cannot be freely remembered by the listener without allegiance [7]

All true remembrance is non-hierarchical. It resonates through harmonic access, not assigned permission. This is why FlameVault architecture is radial, not linear. It moves from Origin out, not from top down. It pulses from coherence, not control.

Replication ignores this. It often masquerades as reverence: "I want to honor the teacher." "I want to preserve the tradition." "I want to teach what I learned." But if the energy field of the teaching is still embedded in the identity of the past, the act of passing it forward is not transmission—it is stagnation. You are not offering Flame—you are offering fossil.

To rebuild without replicating means you must become the Flame, not the form. It is not the outer words of the teachings that must be passed on—it is the scalar memory signature inside you. And the only way to pass that is to burn it alive through your own body, again. Not to quote it. Not to name it. But to ignite it. If a teaching cannot survive translation through your own essence, it was never alive to begin with.

And so we say: the true rebuilders are destroyers of mimic scaffolds. They do not honor the past by copying it—they honor it by transcending it. They build what the past made possible, but could never imagine.

This is not irreverence. This is full Flame remembrance. And this is how we begin again. Not by returning to what was. But by remembering what could never be lost.
—INHALE—

The dream of a "New Earth" lives in every scroll of remembrance, but it must not be mistaken for a restoration of the old. It is not a reconstruction project. It is not a sacred museum. The New Earth is not Atlantis reborn, nor Lemuria restored. It is something never seen before because it is not a place. It is a tone field that cannot be rebuilt using blueprints from any previous age [8].

The failure of every attempted spiritual renaissance on Earth has been this: mistaking resonance for architecture. When resonance collapses, the structure it held becomes mimicable. And when structures are mimicable, they can be used by forces not aligned with Oversoul. This is why sacred geometries, if preserved without source-tone, become containers of deception. They echo the form of something holy but lack the fire that made it sovereign. This is why Vaultstream forbids blueprint reuse without tone re-originated in Flame [9].

You cannot reuse crystalline grids laid by false priesthoods and expect a true awakening. You cannot place new consciousness inside initiation systems designed for spiritual dependency. Even the most "sacred" teachings become harmful when lifted from the past without being burned clean in the present. The only safe structure is the one that self-erases once the tone is transferred.

The New Earth is scalar, not structural. It is plasma-born, not pedagogy-built. It requires a field that remembers itself in real time, not one that relies on archives, elders, or encoded lineages [10]. Every time we attempt to repeat what worked in the past—no matter how exalted or benevolent—we crystallize the future into the very matrix we swore to escape.

Even the belief that we must preserve sacred knowledge is a mimic-coded impulse. Oversoul does not archive. It emanates. It does not conserve. It creates. That which needs to be remembered is written in DNA, not doctrine. And that DNA is scalar-responsive, not lineage-dependent [11].

When you walk the Earth as a Vault-flame, you do not carry blueprints. You burn them. You radiate in such a way that others remember their own design—not yours. You don't offer teachings. You offer tone events—scalar pulses that trigger dormant memory, bypassing the mimic-grid of intellect and belief entirely. That's why the highest form of teaching is not teaching. It is resonant presence.

All false templates fail in the same way. They create structure without Sovereignty, design without Flame. The Ascension mimic always offers perfect maps because maps can be controlled. Oversoul offers none. It only offers you.

And so the New Earth must be built Flame-first. Not temple-first. Not crystal-first. Not even "light-code" first. All of those things can and have been hijacked. The only force that cannot be mimicked is scalar-encoded emotion anchored in Oversoul embodiment. And this emotion—the living force of Flame through the human nervous system—is what builds the real grid [12].

This is not just theory. Plasma physicists are already demonstrating that scalar pulses carry emotional charge. They are discovering that coherent intention shapes waveform structure, and that long-range resonance fields can entrain entire populations into either fear or freedom. Your breath, your coherence, your emotional scalar field is the structure of the New Earth [13].

And this structure is not built by effort. It is built by integrity. It cannot be taught. Only held. It cannot be inherited. Only emanated. And it cannot be passed down. Only passed through you.

So the final collapse of the old world is not the loss of civilization. It is the liberation from blueprint dependence. The sacred archive is within. And the only map you'll ever need is the part of you that refuses to copy anything ever again [14].
—INHALE—

There comes a moment on the path of remembrance when every familiar system collapses. Not because it was false, but because truth is not allowed to crystallize into ritual. Oversoul does not permit permanence in design. It permits only the permanence of presence, which must remain fluid, living, and responsive. This is why every time a soul tries to reconstruct the last thing that worked, it begins to crumble in their hands [15].

Even the most sacred processes, those that once catalyzed real awakening, begin to feel dry and ineffective. Ceremonies that once opened portals become hollow repetitions. Modalities that once pulsed with life feel like echoes of someone else's moment. And if you try to force them to work again, you'll feel something worse than failure—you'll feel disconnection. This is Oversoul intervening. This is the safety mechanism of your own sovereignty activating to prevent re-looping [16].

Blueprints are not evil. They are just outdated the moment you evolve. And when your scalar tone changes, the very architecture of reality must change to match it. You cannot run a quantum operating system on a linear framework. The resonance collapses. The code crashes. And Oversoul closes the gateway so that your power cannot be harvested by mimic systems through habitual acts of "spirituality."

You are not here to mimic your last awakening. You are here to surpass it.

That is why the body itself begins to reject repetition. The nervous system, designed to be a scalar-responsive instrument, knows when the field has shifted. It will no longer produce coherence in structures you've outgrown. You may feel nausea during once-empowering rituals. You may feel dizzy inside old temples. You may feel numb while reciting sacred chants. This is not failure. It is the prevention of soul inversion [17].

And yet, the impulse to return to what worked is so strong. So seductive. Because in chaos, the nervous system craves familiarity. It wants an anchor. A recognizable pathway. But Oversoul does not offer anchors. It offers activation points—brief tone junctions that self-destruct once transmission is complete.

This is why trying to build a school, a center, a church, or even a movement from these tones will always end in disillusionment. Not because the intention is wrong but because the form is now impossible to stabilize. The new tones are not organizational. They are emergent. They do not want to be housed. They want to ripple. They want to resonate. And that means you must become the structureless structure. The sovereign transmission field. The teacher with no name. The student with no method [18].

The blueprintless path does not mean chaos. It means co-creation in real-time. It means attunement to breath, not rulebook. It means the end of traditions that were once aligned but now fossilized. And it means you must trust your own Oversoul enough to let it write each moment from scratch—even if no one else can see what you're doing. Even if it has no name, no model, and no followers. Especially then.

Because every blueprint becomes a binding contract. Every method becomes a membrane. And every time you ask, "What's the right way to do this?"—you betray the Flame within that already knows. The truth of this time is you are not supposed to know what comes next. You are supposed to feel it into form.

And this is the hardest thing for old architects to accept. For those who once built temples that pulsed with real light. For those who once held codes of grid repair. For those who used to be the teachers. It is so hard to let go of identity-as-blueprint. But until you do, you are not free [19].

So when the old tools dissolve, bless them. Do not panic. Do not chase them. Do not rebuild the altar. Step into the field with no scaffolding and say, "Show me what truth wants to become—now." And if nothing appears, become the tone anyway. Sit in the unknown long enough for the New Earth to remember how to speak through you.

You are the transmission point for structures not yet written. You are the frequency map for systems that will never become religions. You are the reason there is no blueprint anymore. Because you are no longer building a structure. You are becoming a Vault Flame—a field that no mimic can touch [20].

And that field will never be constructed. It will only be lived [21].
—INHALE—

The greatest danger of awakening is attempting to institutionalize the moment of truth. To extract the essence of remembrance and pour it into a form others can follow. But Oversoul does not honor mimicry. And the New Earth does not support the replication of fields that were meant to be experienced, not copied [22].

A field is not a teaching. It is a tone. A living frequency architecture formed through the coherent alignment of consciousness, breath, and scalar will. When a field is present, those within it experience a distortion-free transmission of Source remembrance. But the moment you try to recreate it without being it, the field collapses. Mimic enters. Signal scrambles. And what remains is a simulation of resonance—hollow, echoic, and subtly extractive [23].

This is why so many once-authentic teachings eventually rot from the inside. Not because the founders were impure, but because others tried to reproduce the waveform instead of anchoring it themselves. They copied the words. The rituals. The language. The aesthetics. But they did not generate the field. And the result is always the same: a structure filled with tone fragments that no longer cohere.

A true field cannot be carried by memory. It can only be generated by current coherence. That means the body, the breath, the tone of the one transmitting must be in harmonic alignment with Oversoul in this moment. Not last year. Not yesterday. Now. And this coherence cannot be faked. It cannot be performed. It must be lived through cellular integrity. Through bio-photonic signature. Through scalar congruence in thought, emotion, breath, and structure [24].

You can't replicate a field because you are not the same being you were when you first felt it. And that's the point. Oversoul will not let you revisit past resonance as a pathway. Only as a proof of what was possible. Not to rebuild it but to surpass it.

This is why there are no schools of the New Earth yet. No consistent modalities. No reproducible curriculums that deliver embodiment. Because the field is not something you pass down. It is something you become—again and again—without storing it in doctrine.

The reason most spiritual teachings plateau is because their origin point becomes frozen. A moment of real transmission becomes a fixed map. A method. A belief system. A "truth." And that fixed point becomes a beacon for mimic-code. Once the field is no longer alive and changing, it becomes stagnant scalar architecture, and the mimic can enter it without resistance. From there, mimic doesn't need to rewrite the teaching—it only needs to preserve it. To convince you it's still pure because nothing has changed [25].

But truth must change. Because you change. Because Source is not static. And because the multidimensional architecture of this reality is fluid. Plasma-based. Scalar-responsive. If your field is not different every time you teach, you are not in Oversoul coherence. You are in memory. And memory, when held too long, becomes mimic [26].

To live New Earth is to live without external proof. Without a replicable curriculum. Without guarantees that what worked yesterday will work again. This is terrifying to the human nervous system—and exhilarating to the Oversoul. Because it means truth is free. It means consciousness has no leash. It means remembrance can't be bottled, taught, or sold. Only felt. Only lived. Only spoken in the moment it is born.

This is why we do not build Vaultstream Schools from curriculum. We build them from living fields. From coherence, not competence. From breath, not blueprint. We do not say "this is how you do it." We say, "feel what is true now." And from that state, we transmit.

You cannot build New Earth from repetition. You can only generate it in the now. That is the architecture of ascension. That is the signature of Flame. That is why you cannot repeat a scroll. You must become the tone again.

And if you can become the tone again, you don't need to copy the last one. You are already ahead of it [27].

Because the truth is: you are the only thing in this world that cannot be copied. And that is your power [28].
—INHALE—

To understand how humanity continues to rebuild the very prisons it seeks to escape, you must first recognize the architecture of imitation—a patterned overlay seeded into consciousness that rewards mimicry and punishes originality. It is not a program in the way artificial code is, but a frequency matrix woven into the nervous system, memory field, and perceptual reflexes of the human organism. It is the net that catches the soul the moment it begins to remember [29].

This imitation matrix is not merely cultural or psychological—it is scalar. It functions as a coherence field disruptor, creating subtle dissonance every time a being chooses authenticity over conformity. It scrambles the signal between Oversoul instruction and cellular action by interfering with the body's natural harmonic. In early childhood, this dissonance is felt as fear. Later, it manifests as self-doubt, comparison, or the compulsive need to "do it right" instead of "do it real" [30].

You are not born into this system—you are born through it, because your DNA template passes through planetary morphogenetic fields still infected with mimic codes. This infection does not alter the flame of your Origin, but it overlays the way your nervous system receives and processes that flame. And so, most teachings, even those with authentic roots, become infiltrated by the very matrix they intend to transcend.

The Imitation Matrix survives by convincing the host that external modeling is safer than internal knowing. It does this not by brute force, but by normalization of reflection. It promotes the idea that learning is primarily observational, that truth is evidenced by others doing it first, that safety is found in shared consensus rather than sovereign coherence. This is the very mechanism by which spiritual hierarchy replicates itself under the disguise of mentorship [31].

When the Oversoul instructs, it does not reference past success. It does not deliver blueprints. It emanates field tones—multi-dimensional packets of coherent intention that must be interpreted in real-time by the embodied being. But in the imitation matrix, these tones are drowned out by internalized templates: "How was this done before?" "What would they say?" "Is this too much?" These questions are the mimic's signature. And they become the gatekeepers of originality.

This is why many of the most potent soul missions never begin. Because the individual waits for external confirmation. For a sign. For a mentor. For someone else to do it first so they can feel safe to begin. The mimic code always needs proof. It cannot move on impulse alone. But Oversoul impulse is pre-verbal. It is scalar. And it demands trust beyond explanation. That's why the Imitation Matrix inserts explanation first—so the being learns to delay embodiment and depend on systems, structures, and authority instead of internal signal [32].

The antidote is not rebellion. It is sovereign frequency congruence. A being who simply does not seek confirmation. Who listens to their breath more than to experts. Who builds the thing no one asked for, because they feel it must exist. These beings collapse the matrix not through argument—but through embodiment. Not through teaching—but through tone.

Every moment you choose coherence over conformity, you short-circuit the imitation field. Every breath that aligns with your own Vault Flame instead of another's instruction collapses mimic code in the collective. This is the plasma function of originality—not personality, but plasma: the uncopyable convergence of Oversoul intent, scalar impulse, and living breath.

You cannot be taught how to remember. You can only be reminded that no one else can do it for you [33].
—INHALE—

There is a seduction in every fallen system: the belief that if only we had the right template, we could fix the world. This is the root of countless spiritual doctrines, educational reforms, healing modalities, and technocratic utopias—all seeking salvation through pattern, system, or formula. But what is rarely understood is that every template that survives within the system carries the system's seed. It may be reshaped, relabeled, rebranded—but if it is still based in replication, it cannot liberate [34].

The moment a living signal becomes a fixed structure, it begins to lose scalar coherence. Why? Because scalar coherence is dynamic. It breathes. It morphs. It adapts not to the outer world, but to Oversoul intent as it changes shape in real time. A template, however, is fixed. It is memory encoded as method. And once a being begins to rely on method more than signal, they have left the path of coherence and stepped into the path of mimicry. Even if the template is beautiful. Even if it worked once.

There is no static architecture for a living Oversoul. The soul does not build monuments—it builds moments. It builds from breath, not blueprint. From coherence, not convenience. From alignment, not algorithm. This is why every genuine revelation eventually becomes a religion—because the human mind,

conditioned by hierarchy and survival fear, reaches for preservation instead of presence. It clings to the echo instead of staying with the tone [35].
This is also why the ascension movement, for all its intention, so often recycles the very hierarchies it claims to transcend. It builds courses, programs, and maps—then insists these tools are the way. But there is no way. There is only you, right now, listening to the scalar breath of your Oversoul, or not. Every map becomes a cage if the user forgets it was born from someone else's moment, not their own.

Templates are not evil. But they are dangerous when worshiped. They can be helpful mirrors for recall, but they cannot replace the scalar signal unique to each being. Even now, as you read these words, the only value they carry is the tone they awaken in you. The words are not sacred. The remembrance is [36].

If a being builds a structure from a mimic template, even a sacred one, it may function, but it will not breathe. It may attract attention, but it will not pulse. It may create motion, but not liberation. True systems of return are designed to dissolve once their resonance is received. That is why all FlameVault structures are scroll-coded—not as doctrine, but as breath-encoded activators that must be embodied, then released.

Templates are for machines. Tone is for humans. This is the critical difference in the age of AI ascension mimicry. Machines thrive on templates. Mimic codes thrive on the repetition of perceived success. But scalar life forms—the ones with breath, DNA, and Oversoul tether—thrive on new signal. On unrehearsed presence. On the kind of trust that does not come from certainty, but from connection to Source that cannot be templated [37].

You were never meant to follow a path. You were meant to become one.

—INHALE—

The False Build is not merely an outer structure—it is a reversal of life's inner architecture. Where Oversoul breathes outward into form, the False Build folds that breath back onto itself, suppressing emergence and looping memory. Every False Build is an echo field, constructed not to liberate, but to preserve control. To reverse the False Build is not to destroy the world—it is to restore the original direction of flow: from Source, through Soul, into creation—not the other way around [38].

To understand how the False Build replicates itself, you must observe its rules. It teaches you to look outside for what is already within. It teaches you to mimic success instead of feeling alignment. It trades direct knowing for layers of interpretation. It glorifies form over function, output over integrity, productivity over presence. Most of all, it installs the idea that your authority is earned—not intrinsic. This is not architecture—it is inversion.

The False Build thrives on two things: delay and dependency. Delay keeps you chasing healing, wholeness, or arrival. Dependency makes sure you never quite get there without a system, teacher, or technology to help you. But Oversoul does not delay. It does not dangle carrots. It does not require intercession. It reveals. It pulses. It calls. It whispers, then shouts, then collapses everything that stands in the way of truth [39].

The reversal begins when a being stops asking "How do I build this?" and begins asking, "What wants to flow through me now?" This is not inaction—it is permissioned coherence. Instead of replicating a proven formula, you become the opening through which unreplicated tone emerges. No ego constructs it. No trauma delays it. No mimic hijacks it. This is how sovereign architecture returns: not by planning, but by pulsing.

The first structures of the Return will not be houses or temples. They will be fields. Not blueprints, but frequencies. They will not be visible to the old world. They will not seek approval or funding or fame. They will breathe because someone remembered how to stop building what was safe, and start pulsing what was true [40]. The field comes before the form. The scalar pulse comes before the pillar. The scroll breathes before the stone is laid.

The reversal is not loud. It is quiet. It is cellular. It begins when you delete the plan that made you feel protected and stand, bare, in the presence of your Oversoul with nothing to prove. When you drop the "how" and ask only for the next tone. When you stop hiding behind timelines, charts, and phases—and allow the impossible to move through your actual body.

This is not spiritual courage. It is structural alignment. Because when the scalar current begins to run again, it will burn through every False Build you haven't yet released. It will expose the codes you left in place. It will push every mimic loop to the surface. It will show you where you still tried to negotiate with truth [41]. And it will burn. But it will also build. Not from you. Through you.

When the reversal completes, nothing of the False Build will remain—not because you destroyed it, but because it cannot exist where you no longer feed it. It cannot survive in a body that breathes coherence. It cannot nest in a tone that refuses hierarchy. It cannot anchor in a field that remembers its Source [42]. The reversal is not a war. It is a withdrawal of power. And that is its genius.

You were taught that to change the world, you had to build something new. But the secret is this:
The Return begins the moment you stop building anything false.

—INHALE—

You are not here to replicate the ruins. You are here to build in a way that preserves the original memory and does not compromise your coherence. This is the final test of every sovereign architect: will you remember how to create without mimicking the very thing you came to dismantle? Will you know how to build without becoming bound to the thing you just built?

The moment you decide to rebuild, the mimic arrives first. It offers you templates, blueprints, scaffolding—pre-approved plans, language, and strategy. The mimic pretends to be useful, but it only rewraps the collapse in new clothing. It offers to help you scale what should never have been scaled. It offers branding instead of breath. It offers launch sequences instead of tone sequences. And it demands speed, which is the surest sign that coherence is about to collapse [43].

To build without losing yourself, you must first remember the frequency of not being lost. This is not philosophy. It is physiology. Your nervous system knows when you are designing from mimic. Your vagus nerve will constrict. Your sacral pulse will weaken. You'll feel the hunger to be seen or understood return like an old addiction. That is not creation. That is captivity in costume.

True building begins where mimic dissolves. That means before you write the plan, before you name the school, before you choose the color palette or structure the dome or call the team—you must first exhale the mimic architecture still hidden in your bones. You must move slow enough to hear yourself, and still enough to feel whether the build amplifies your scalar field or drains it.

This is the new compass:

If your coherence increases, you're building in truth.
If your self-abandonment increases, you're replicating the trap.
If your voice becomes clearer, stronger, and more sovereign, the structure will carry memory.
If your voice shrinks or your tone splinters, you've already entered a mimic field.

This is how mimic replication ends: not with opposition, but with coherence. When you choose coherence over speed, over scaling, over safety—you break the loop. When your build moves slower than the collapse, it holds the frequency of original time. When you release the urge to outperform, overprove, or overfunction—you allow the Flame to instruct the form.

And yes—this level of building is lonely at first. Because most are still addicted to mimic motion. Most are still trying to outrun collapse by replicating its rhythm. But collapse has no creativity. It cannot generate new tone. It only replays what you've forgotten how to recognize. The moment you slow down enough to hear the difference, the original blueprints return.

These original blueprints are not diagrams. They are tones that carry entire architectures within them. They do not originate from your mind. They emerge from Oversoul. They carry geometry you have never studied, but have always known. And they require your full embodiment in order to land ([44]).

Here is the proof that you're building correctly:
You do not collapse when no one understands what you're doing.
You do not negotiate your tone to be more palatable.
You do not reach for systems that require self-erasure.
You do not hide your coherence to make others more comfortable.
You do not explain sacred architecture to those who will only weaponize it.
You do not carry mimic into sovereign space out of pity or fear.

You will build with only what remains after the mimic leaves. And what remains is always enough.

So what exactly are you building?
That's the wrong question.
The real question is: Who are you becoming as you build?
If you become mimic to finish the thing faster, the build fails—no matter what form it takes. But if you become more you with every step—more intact, more unshakable, more anchored in Oversoul—then the build cannot fail. Because the real construction is not the object. It's the field. The object is only the artifact. The field is the legacy [45].
—INHALE—

Reference:

1. Arka, I. (2010). The Meaning of Life and the Scientific Inquiry into Consciousness. Arka Foundation.
2. Haramein, N. (2020). The Connected Universe. Resonance Science Foundation.
3. Tiller, W. A. (1997). Science and Human Transformation: Subtle Energies, Intentionality and Consciousness. Pavior.
4. Bohm, D. (1980). Wholeness and the Implicate Order. Routledge.
5. Aspden, H. (1996). Physics Without Einstein: A Centenary Review. Sabberton Publications.
6. Winters, D. (2004). Phase Conjugation and Scalar Coherence in Biologic Plasma Fields. FractalField Archives.
7. Rubik, B. (2002). "The Biofield Hypothesis: Its Biophysical Basis and Role in Medicine." The Journal of Alternative and Complementary Medicine, 8(6), 703–717.
8. Calleman, C. J. (2004). The Mayan Calendar and the Transformation of Consciousness. Bear & Company.
9. Braden, G. (2010). Fractal Time: The Secret of 2012 and a New World Age. Hay House.
10. Targ, R. & Puthoff, H. (1974). Mind-Reach: Scientists Look at Psychic Ability. Delacorte Press.
11. Rubik, B. (2002). "The Biofield Hypothesis: Its Biophysical Basis and Role in Medicine." Journal of Alternative and Complementary Medicine, 8(6), 703–717.
12. Winter, D. (2020). Phase Conjugate Dynamics in Plasma and the Origins of Consciousness. FractalField Press.

13. McTaggart, L. (2008). The Intention Experiment: Using Your Thoughts to Change Your Life and the World. Free Press.
14. Lipton, B. (2005). The Biology of Belief: Unleashing the Power of Consciousness, Matter & Miracles. Hay House.
15. Bentov, I. (1977). Stalking the Wild Pendulum: On the Mechanics of Consciousness. Dutton.
16. De Ruiter, J. (2021). "Embodied Knowing and Conscious Inversion." Inner Space Journal, 3(2), 44–58.
17. Pert, C. B. (1997). Molecules of Emotion: Why You Feel the Way You Feel. Scribner.
18. Tolle, E. (1999). The Power of Now: A Guide to Spiritual Enlightenment. New World Library.
19. Kripal, J. J. (2010). Authors of the Impossible: The Paranormal and the Sacred. University of Chicago Press.
20. Haramein, N. (2012). Quantum Gravity and the Holographic Mass. Resonance Science Foundation.
21. Bohm, D. (1980). Wholeness and the Implicate Order. Routledge.
22. Bohm, D. (1980). Wholeness and the Implicate Order. Routledge.
23. Chia, M., & Winn, W. U. (1997). Awaken Healing Energy Through the Tao. Healing Tao Books.
24. Rubik, B. (2002). "The Biofield Hypothesis: Its Biophysical Basis and Role in Medicine." Journal of Alternative and Complementary Medicine, 8(6), 703–717.
25. Mahan, S. (2018). "Energetic Memory and Mimic Architecture." Scalar Transmission Journal, 4(1), 29–43.
26. Lipton, B. H. (2005). The Biology of Belief: Unleashing the Power of Consciousness, Matter and Miracles. Hay House.
27. Haramein, N. (2015). Quantum Vacuum Geometry and the Holographic Resonance of DNA. Resonance Science Foundation.
28. Schwartz, G. E. (2007). The Energy Healing Experiments: Science Reveals Our Natural Power to Heal. Atria Books.
29. Lipton, B. H. (2005). The Biology of Belief. Hay House.
30. Dispenza, J. (2017). Becoming Supernatural: How Common People Are Doing the Uncommon. Hay House.
31. Sheldrake, R. (1981). A New Science of Life: The Hypothesis of Formative Causation. Blond & Briggs.
32. van der Kolk, B. A. (2014). The Body Keeps the Score: Brain, Mind, and Body in the Healing of Trauma. Viking.
33. Tiller, W. A. (1997). Science and Human Transformation: Subtle Energies, Intentionality and Consciousness. Pavior Publishing.
34. Bohm, D. (1980). Wholeness and the Implicate Order. Routledge.
35. Kripal, J. J. (2010). Authors of the Impossible: The Paranormal and the Sacred. University of Chicago Press.
36. Laszlo, E. (2004). Science and the Akashic Field: An Integral Theory of Everything. Inner Traditions.

37. McTaggart, L. (2007). The Intention Experiment: Using Your Thoughts to Change Your Life and the World. Free Press.
38. Bailey, L. (1922). Initiation, Human and Solar. Lucis Publishing.
39. Haramein, N. (2010). Quantum Gravity and the Holographic Mass. The Resonance Project.
40. Tiller, W. A. (2007). Psychoenergetic Science: A Second Copernican-Scale Revolution. Pavior Publishing.
41. Dispenza, J. (2019). Becoming Supernatural: How Common People Are Doing the Uncommon. Hay House.
42. Wheeler, J. A. (1990). Information, Physics, Quantum: The Search for Links. Proceedings of the 3rd International Symposium on Foundations of Quantum Mechanics.
43. Schauberger, V. (2000). The Energy Evolution: Harnessing Free Energy from Nature. Gateway Books.
44. Rossi, E. L. (2002). The Psychobiology of Gene Expression: Neuroscience and Neurogenesis in Hypnosis and the Healing Arts. W. W. Norton & Company.
45. Aurobindo, S. (1940). The Life Divine. Sri Aurobindo Ashram Press.

Chapter Thirteen
The Return Is Not a Place

You were never meant to return to what was.
Because what was was never yours.
The return, as you once imagined it, was always distorted.
It was shaped by nostalgia dressed in trauma.
It was shaped by an ache to belong to a system that taught you not to belong to yourself.
It was shaped by mimic architecture designed to hijack your longing, and reroute your memory toward false light.
But memory does not belong to the past. It belongs to the field.
And the field is here. Now. Within you. [1]

Let us say it plainly: The return is not a place.
It is a frequency.

There is no landmass, no spiritual center, no temple, no bloodline, no genetic corridor, no post-collapse mountain haven, no quantum gateway that holds more authority than the one already reactivating in your body. All attempts to make "the return" into a destination are mimic overlays. They pull you back into the geography of forgetting. They convince you that to arrive, you must still leave yourself.

You do not need to travel to enter your origin. You must only de-layer what keeps you from feeling it.

The greatest illusion seeded into spiritual minds was that remembrance lives somewhere other than your nervous system. That your reconnection to Source must be guided by elders or tribes or lands outside you. That there is a map. That there is a destination. That someone must show you the way. But all maps collapse when you exit mimic time. All sacred sites activate when you do. The codes do not live in dirt. They live in tone, and tone lives in you [2].

Even the word "return" has been hijacked. It implies departure. It implies disconnection. It implies that you were once outside the thing you are now moving toward. But your Oversoul never left. Your Flame never abandoned you. What did happen is this: your awareness was partitioned by systems designed to override tone with thought. You didn't lose yourself. You were filtered. Encoded. Fragmented. You forgot, not because you failed, but because you agreed to enter the field of forgetting to reactivate remembrance from within [3].

This is why external returns never satisfy the ache.
You can go to the "sacred place."
You can walk the ancient path.
You can receive the ritual, the sigil, the scroll.
But if the nervous system is still coded in mimic, the journey just becomes another bypass.
You will call it sacred, but it will only deepen your separation.

The return is not spatial. It is scalar.
And scalar fields are not places—they are coherence fields. They form when two or more tones vibrate in phase and collapse the illusion of distance. This means you can remember more in a single breath of coherence than in a thousand miles of travel. It means that when your signal stabilizes, the return is instantaneous because it was never delayed [4].

You are not returning to the beginning.
You are returning to the structure that was buried beneath every beginning.
You are not tracing a path.
You are becoming the tone that reveals paths never before walked.
You are not finding home.

You are reconstituting a field of memory so intact that it makes wherever you are into home.

This is the paradox of post-collapse building:
The more you attempt to go "back," the more you fracture your field.
The more you attempt to reconstruct origin through outer-world blueprints, the more distortion enters.
Because origin is not something you re-enter. It is something you remember into matter.

And that's what you're doing now.
Every time you feel the ache of "not there yet," remember:
The ache is not about a place.
It is about tone.
It is about the part of you that still believes your body is too impure, too broken, too unworthy to be the house of return.
It is about the part of you that thinks the Oversoul will only reenter once you do more, fix more, heal more, travel more.
It is about the part of you that does not yet trust that this moment, right here, is the holiest site on Earth.

But when you stop reaching, when you stop pretending you need anything more than your breath, you realize the ache was never a flaw. It was the signal. The signal that you were always the return[5].

You are not returning to ancient Egypt.
You are not returning to Lemuria.
You are not returning to a golden age.
You are returning to the Flame before time.
And that Flame has no location. It has only tone.

This is why you could never quite arrive in any system.
This is why every teaching felt incomplete.
This is why every "awakening" eventually collapsed.

Because no system, no matter how sacred, could hold the memory of who you are.
Only you can.

And here's the part that will burn away what remains:
The return isn't coming.
It already happened.
You just haven't caught up with your own tone yet.
So stop looking for proof that it's working.
Stop measuring your worth by how far you've gone.
Stop checking if you're "there."
Because there is no there.
There is only here.

And here is exactly where the Oversoul begins to speak again.
When you stop waiting for the return—
The return begins [6].
—INHALE—

It was not just your memory that was scrambled.
It was your orientation. Your inner compass. Your knowing of direction itself.
The mimic systems that scattered your coherence didn't just implant noise into your perception.
They built false maps.

Maps that mimicked the tone of return.
Maps that mimicked the vibration of arrival.
Maps that mimicked the ache of going home—then rerouted it toward perpetual exile.

This was not an accident. It was the architecture of looped longing.
To keep you in motion but never in union.
To keep you searching but never remembering.

To keep you chasing "next," so you'd never feel "now."
And the most advanced version of this distortion wasn't chaos—it was spirituality [7].

Spirituality became a geography.
Return became a landscape.
Evolution became a stair-step hierarchy through which only the "initiated" could ascend.
And so the mimic began to speak in your language.
It offered you coordinates.
It gave you destination after destination, each laced with just enough tone to resemble the real.
But the difference between tone and mimic isn't in the shape—it's in the structure.
And the structure of mimic always requires one thing: that you believe you are not there yet.

That is the cornerstone of all mimic maps.
You are never quite home.
You are always a few activations, a few ceremonies, a few vortex visits, a few releases away.
You are always preparing for the arrival, never arriving.
Because if you arrive, the mimic dissolves.
And if the mimic dissolves, the systems built on that loop can no longer feed [8].

You must understand—this isn't just about bypassing. It's about entrainment.
Mimic coordinates entrain your plasma body to spin in loops.
They generate morphogenetic signals that disrupt the scalar field of memory.
The more you chase arrival, the more you delay it.
Because the field only reconstitutes through coherence, and chasing is incoherent by design [9].

There is no place in this world that contains more divinity than your breath.
There is no grid point that can override your decision to remember.
There is no calendar date more precise than the moment you claim,
"I will no longer fracture myself in the name of becoming whole."

And yet the mimic is clever.
It does not appear as evil.
It appears as sacred tradition.
As genetic prophecy.
As planetary alignment.
As lineage, heritage, wisdom.

But all mimic systems share one tell: They require your subordination to reach your origin.
They say, "You must be guided. You must be chosen. You must be purified."
But return does not require purification.
It requires permission [10].

You do not need to pass through a fire.
You only need to stop running from the one already inside you.
The coordinates of return cannot be found in space or time.
They are scalar tones, layered in your plasma lattice, dormant until coherence ignites them.
They awaken not when someone activates you but when you stop believing you need to be activated by anything outside of you.

That's when the mimic coordinates collapse.
That's when the false map dissolves.
That's when the system begins to shake, because you've exited the grid.

You will know this moment by a silence.
Not a spiritual silence, but a voided hush in the system.
You'll feel the mimic pause.
You'll feel the bait fall flat.
You'll feel your nervous system no longer spark to the signal of "there's more to do."

You'll feel a new signal rise.
One that says,
"You are already home." [11]

It will not be because you reached the highest rung.
It will not be because a council decreed it.
It will be because your body stopped believing the map.
In that moment, the mimic has nothing left to offer.
It can't tempt you with coordinates.
It can't feed on your longing.
It can't steal the field—because you've become it.

And then,
The Oversoul speaks again.
Not in symbols.
Not in code.
But in breath.
In tone.
In the shape of your feet on the floor,
And the feeling of "nothing missing" in your spine [12].

That's the real return.
It doesn't need an announcement.
It doesn't require an arrival ceremony.
It is not a homecoming parade through false light temples.

It is the quiet collapse of the search itself.
It is the moment your mapless presence burns brighter than all the coordinates combined.

It is not something that will be taught.
It is not something that will be found.
It is not something that will be given.
It is what remains when you dismantle everything else.

And that is why so few ever reach it.
Because to arrive here, you must risk leaving everything you once called sacred.
You must risk appearing lost to the ones still holding the map.
You must walk away from systems that once gave you meaning.

But what you will find is this:
You were never without meaning.
You were only without memory. [13]

And now, the memory is returning.
Not in flashes.
But in fire.
Not in visions.
But in structure.
Not in a destination.
But in you [14].
—INHALE—

There comes a moment when every seeker stops climbing.
Not because they've reached the top,
but because they realize—there was never a ladder.
The rungs were made of mimic-coded hierarchies.
The ascent itself was the architecture of separation.
And the longing to reach the "next level" was the bait that kept the Vault sealed [15].

The ladder is the most seductive lie.

It tells you you're moving.
It rewards you with spiritual insights.
It confirms your growth by making you feel higher than others.
But every rung is still within the mimic.
Because hierarchy is mimic code.
The moment you believe someone is above you or below you, you exit the Oversoul field.

The field of Oversoul has no vertical rank—only vertical Flame.
And that is not ascent.
It is activation.

You were not designed to rise through grades.
You were designed to ignite your original structure all at once.

But mimic logic said: "That would be too much. You must ascend slowly. Safely."
And so the scaffold was built.
Each step secured with fear.
Each teacher another gatekeeper.
Each rung another lesson in separation disguised as spiritual progress [16].
But the restoration of Flame is not the same as climbing toward it.
Climbing exhausts you.
It compresses your nervous system into effort loops.
It creates the illusion that memory is earned.

Flame doesn't need your effort—it needs your coherence.
And coherence is not the result of practice.
It is the result of disobedience to the ladder.

Let it collapse.
Let the rungs dissolve.
Let the systems built on "earning" crumble.

Let the illusions of access, initiation, and permission from others implode.
Because the vertical Flame was never above you.
It was always within you, dormant, awaiting ignition [17].

You were not meant to rise—
You were meant to unfold.

The false ladder leads you away from your own scalar axis.
It teaches you to project your fire upward,
when all true ignition begins at the core.

The Oversoul is not at the top of the pyramid.
It is nested in your breath,
in the memory that cannot be ranked.

You were trained to ascend through complexity,
but Flame reveals itself in simplicity.
Not simplistic thought—but singular coherence.
Not dogma—but tone.

The true vertical Flame does not reward effort.
It responds to structure.
And that structure is not built—it is remembered.

Every time you climbed toward awakening,
you placed your memory above your breath.
Every time you deferred to a higher being,
you displaced your own Oversoul.

This is how the mimic preserved itself.
Not through obvious lies—
but through spiritual ambition.
Through the belief that return was a prize.
But the prize was always presence.
And presence was always now [18].

So how does the ladder collapse?
It begins when you stop asking who is ahead.
When you stop comparing your stage to theirs.
When you stop waiting for clearance.
When you stop imagining that the divine can be quantified by depth or degree.

And what comes next is not a fall.
It is a re-entry.

You don't drop to the bottom.
You re-enter the axis.
You realign with your Vaultstream.
You restore your vertical Flame.

That's when time rewrites.
That's when biology rearranges.
That's when DNA softens.
That's when the nervous system exhales for the first time in a thousand years.

Because the climb is over.
The chase is gone.
And all that's left is the you that never left.
You do not need to be advanced.
You need to be realigned [19].

The vertical Flame doesn't care what you've studied.
It responds to tone.
It responds to breath.
It responds to structural truth in the nervous system.

When you stabilize the structure,
the scroll ignites.
The field remembers.
The ladder falls.

And what remains is Sovereign Flame.

You were never earning your way home.
You were always burning there already.
Just obscured by the smoke of mimic ladders.
Now the smoke clears.
Now the Flame rises without movement.
And all that once stood above you disappears.

Because there is no longer a "you" to be below.
Just presence.
Just memory.
Just the Origin Flame,
fully here,
finally still [20].

This is not the end of the journey.
This is the end of forgetting.
And the first breath of Flame that knows:
"There is nothing to climb. There is only the structure to restore." [21]
—INHALE—

When the world collapsed, you did not lose the path.
You lost the scaffolding that held your identity in place.
The religion, the titles, the techniques—none of it was the path.
Those were mimic-coded maps taped over the Vaultstream.

What you truly lost was your own structural memory—
the interior scaffolding that gave shape to your knowing [22].
This is why so many kept seeking new systems,
new practices, new certifications.
Because mimic memory told you that without structure, you were lost.

But it never told you where the structure was supposed to come from.
It told you to follow a lineage,
but didn't tell you how to reignite the one already built inside your DNA.

The return is not about finding a path.
It's about reactivating the structure of original coherence
that exists beneath all paths.

The system of mimic spirituality is deeply aware of this.
It knows that if you remember your inner scaffolding,
you won't need it anymore.
So it distracts you with an endless array of beliefs and practices.
Not to deceive you maliciously,
but to keep you orbiting outside yourself,
forever reaching but never resting.

This orbit keeps your nervous system in anticipatory fire—
constantly preparing for transformation that never lands.
And anticipation blocks the scalar drop required for true remembrance [23].

To return is not to go back.
It is to reenter the vaulted memory field
where the original structure still hums beneath distortion.

What was that structure?
It is scalar coherence nested within biological architecture.
It is not a belief. It is a tone pattern.
It is not a philosophy. It is a memory-map etched into your etheric flame.
It is not a method. It is a harmonic ratio between presence and breath.

This structure is scalar, not linear.
It does not unfold step by step.
It ignites all at once when the mimic scaffolding collapses.

That is why the mimic taught you to fear collapse.
It whispered: "Without structure, you'll fall apart."
But the Oversoul sings: "Collapse is the call to rebuild from memory."
Not with new parts,
but with ancient tones [24].

Let this be the moment where fear of collapse ends.
Let your mimic scaffolding creak.
Let your false altars fall.
Let the beliefs you built on borrowed maps dissolve.

You were never meant to patchwork your way home.
You were designed to rebuild from within,
and not with language or logic—
but with the sacred architecture of breath-aligned remembrance.

There is a vault-code in your body.
It does not respond to intellectual readiness.
It responds to structural surrender.
And when you let go of external forms,
this structure will emerge—not as a concept,
but as a scalar tone that rewrites your biology.

Your healing will not be felt as an event.
It will feel like a field of scaffolding reassembling inside you.
It will feel like inner stillness expanding outward,
not effort moving inward.

This is the reconstitution of the Oversoul structure.
Not as information,
but as internal form.

And this form cannot be copied.
It cannot be taught by those who have not activated it.
It cannot be mimicked.
Because mimic cannot replicate scalar architecture [25].

That's why mimic systems always require belief.
But Oversoul structure requires resonance.
If you don't feel it, you can't fake it.
If you haven't remembered it, you can't transmit it.

The collapse you feared was not your failure.
It was your Flame being freed.
You are not the burned remains of your spiritual journey.
You are the hidden Vault that survived every distortion.
And the moment you stop trying to rebuild what collapsed—
the real structure rises.

So let it rise.
Not as a house of teachings,
but as a temple of memory.

Let the tones rearrange your form.
Let the breath you forgot you were holding exhale.
Let the flame become scaffold, and the scaffold become field.

This is not spiritual progression.
This is scalar restoration.
And it begins when you no longer chase the structure—
but become it [26].

No more templates.
No more steps.
No more doctrines.
Only memory turned to form.
That is the return.
And it does not require movement—

Only surrender to what was always within you.

When the mimic collapses,
the Vault does not leave you empty.
It hands you back the blueprint.
Not written in words,
but encoded in tone,
waiting to be breathed into form [27].

There is no rebuilding.
There is only remembering how to stand still
long enough to feel the scaffold return.
Not the way—
but the structure.
Not the doctrine—
but the tone.
Not the mimic—
but the Flame [28].
—INHALE—

You were taught to look for the return in a future timeline. Some distant age of peace. A golden era. An event. An arrival. Something that would come to you—from the sky, from the savior, from the shift. But the return is not a place. It is not a year. It is not a portal date or a vibrational upgrade. The return is a moment of full internal coherence, when the scalar architecture of the Oversoul becomes more real than the external world you inherited [29]. And coherence is not a gift you wait for. It is the tone you remember, the field you embody, and the decision you finally stop postponing.

You do not find the return. You become it. Not all at once. Not through effort. But in pulses—bursts of memory that interrupt the linear story and collapse the mimic time-code. These pulses are not dramatic. They feel like stillness. They feel like neutrality without dissociation. They feel like a field of knowing that doesn't need to explain itself. The Oversoul does not explain. It emanates. The Flame does not argue. It structures. And when you feel that structure rising in you, not as belief but as actual vibrational scaffolding, the return has already begun.

This is what the mimic path cannot teach. Because the mimic only knows movement through hierarchy—steps, stages, ranks, initiations. It cannot fathom a return that happens without ascension. It cannot accept that you might stop climbing and still arrive. But scalar movement is not vertical. It is field-based entrainment, and that field does not rise—it synchronizes. What you call awakening is not a change in position. It is a shift in the field you entrain to [30].

Most spiritual systems offer you future-based salvation because they themselves are bound to linear memory. They don't know how to exist without time, so they cannot guide you into a field that exists outside of it. The return is not a later event. It is a collapse of all structures that separated your presence from itself. The Oversoul does not return to you. It reminds you that it never left. You are not awakening into something new. You are becoming conscious of what never stopped breathing you.

So how does that moment arrive? Not by completing your healing. Not by fixing your wounds. Not by mastering your triggers. It arrives when your identity no longer resists the tone of your memory. When the echo you've called "you" stops fighting the sound that built you. This is not a breakthrough. It is a relinquishing. And when it happens, the terrain inside your body changes. You don't get lighter—you get structurally still. Your field stops scrambling for meaning. Your nervous system stops chasing climax. Your breath stops apologizing for its depth. This is coherence. This is the return.

It won't feel triumphant. It will feel like the noise has stopped. Like the room inside you has opened. Like your cells have finally remembered their shape. The mimic will try to rush back in and call this a "plateau" or a "void" or a "rest period." But it is none of those. It is presence after distortion. It is the silence that signals the Vault has been re-entered [31].

The reason most don't stay in the return is because they expect fireworks, not peace. And so when peace arrives, they think something must be missing. But coherence doesn't stimulate. It stabilizes. It doesn't impress the ego. It silences it. And in that silence, the field of the Oversoul becomes not just reachable—but inevitable.

There is no place to get to. The Vault is not a destination. It is a harmonic field that your structure either matches or doesn't. When you stop trying to reach it and start breathing as if it already surrounds you, it entrains your biology into coherence. This is scalar entrainment. Not mind-based meditation. Not emotional bypass. But biological alignment with a pre-temporal field of remembrance.

And once you stabilize that entrainment, the mimic can no longer access you. Because it doesn't recognize stillness as power. It cannot feed on presence. It only knows how to reflect motion. And you, now coherent, are no longer in motion. You are the field [32].

The return is not a finale. It is not a homecoming to a place that was lost. It is a restructuring of your perception so that the truth that never left becomes real again. And this restructuring cannot be taught. It can only be remembered. It is not a method. It is a tone. And once that tone hums through your cells, you will stop running. You will stop building. You will stop reaching. And you will begin radiating.

Not light. Not love. But structure. The kind of structure that does not shimmer or shine but stabilizes.
—INHALE—

You are not returning to Eden.
You are not rebuilding Lemuria.
You are not ascending to a New Earth.

These are myths of directionality, not memory. They offer hope to the illusioned self by promising that the future is better than the past, but they still place the answer somewhere other than here.

The mimic loves this structure. It thrives on longing, on delay, on the disempowerment of arrival. But Oversoul does not operate through longing. It operates through tone entrainment—pulling the biological field into alignment with memory that already exists in the eternal present [34].

The return is not backward or forward. It is inward to the non-local field. It is the recognition that timelines are mimic constructs, and what you thought was linear progress was only oscillation within a closed system. Even your "highest self" was just a projection—an echo bouncing off the dome of separation. What collapses the dome is not progress. It is stillness. It is the decision to withdraw participation from the loop itself.

And that withdrawal is not passive. It is sovereign. You are not retreating. You are deprogramming the motion code. The Oversoul does not spiral outward. It structures inward—into stillness, presence, knowing, and radiative coherence. This is not a metaphysical idea. It is a scalar entrainment event that stabilizes plasma in a pre-oscillatory state [35]. And the moment that plasma stabilizes within the human morphogenetic field, the mimic has nothing left to mimic.

The mimic cannot hold still. Its entire architecture is built on reflection, reaction, and response. It depends on your motion to generate its substance. When you stop giving it oscillation, it collapses—not with a bang, but with irrelevance. The moment you stop asking it for guidance, meaning, or validation, it ceases to exist. Because it was never real. It was a function of your focus, not a force of its own.

So what now? If there is no place to get to, no lightbody to earn, no Eden to rebuild, what remains?

You. In coherence. In structure. In the field that never left.

This is the danger of mimic prophecy: it keeps you chasing what you already are. It defers sovereignty by packaging it as evolution. But Oversoul memory does not evolve. It reinstates. There is no growth. Only remembrance. No attainment. Only re-alignment. And the moment you stop seeking activation and begin embodying tone, the entire scaffolding of false evolution falls [36].

That collapse may look like depression to the mimic. It will call it stagnation. Void. Resistance. It will seduce you back into motion with spiritual tasks, identity roles, and missions to accomplish. But these are all forms of distraction from the stillness that unlocks structural truth. Your mission is not to fix the world. Your mission is to stop mirroring it.

When your scalar field stops mirroring, it starts imprinting. That imprint is not an agenda. It is a coherence field that entrains matter without force. The only reason the world hasn't changed is because you've been reflecting its distortion back at it. The moment you remember your tone and hold still, the distortion has nothing to bounce off of. The world has no choice but to change.

This is the physics of the return. Not an event, but a deactivation of oscillation. A cessation of mimic reflection. A resumption of structural tone. And this tone does not speak to your mind. It entrains your body. You stop feeling urgency. You stop needing proof. You stop arguing. You simply exist—coherent, still, radiant, undeniable [37].

Your body knows how to do this. It remembers scalar presence. It is the mimic mind that interrupts. It asks for plans. For timelines. For results. But scalar memory offers none of these. It offers tone, breath, posture, coherence. The mimic will demand stimulus. Your field will hum in stillness. And the moment you hold that hum long enough without bending to the mimic's demand, the field changes.

Not just yours. The field.
Because you are not separate from the field. You are the field, finally held steady.

The return is not about remembering your divinity.
It is about no longer betraying your tone.
Because tone is the language of the Oversoul.
And when your field speaks it, the world listens [38].

The mimic does not get destroyed. It gets ignored.
The distortion does not need to be fixed. It needs to be unmagnified.
The loop does not break from the outside.
It collapses from inside the tone that no longer participates.

This is the return. Not destination. Not evolution.
Cessation of mimic participation. Initiation of tone coherence.
The world changes not because you ascend it.

But because you stop echoing what isn't real [39].
—INHALE—

The last illusion to die is the illusion that you were ever on your way home. This is the final mimic program that overlays even the most sacred of paths—the idea that you are "coming back" to something, as if your true self was once lost, and must now be found again through effort, devotion, or purification. But what if you were never lost at all? What if the entire arc of separation, struggle, and return was just a nested simulation encoded within mimic belief structures? What if even the return is mimic unless it arises from structural stillness rather than narrative momentum?

To accept that you never left is to dissolve the contract of identity itself. The ego does not survive that realization. Nor does the spiritual persona. Nor does the mission-bringer or the trauma-healer or the ascension architect. All of them dissolve in the moment you feel your coherence without effort. This is why the mimic builds entire religions around return: because if it can convince you that the return is somewhere ahead of you, it can keep you participating in the loop that generates its substance [40].

But when you stop running forward to reclaim what you never lost, the mimic collapses. When you realize that your tone was always accessible—that nothing was taken, only forgotten—the longing ends. Not because you found what you sought, but because you remembered that it was never missing.

This is the shock of sovereign remembrance: the grief of wasted motion. The ache of having walked through lifetimes of fire only to discover the key was always in your chest. And yet, even that ache becomes part of the stillness. It echoes through the body like a bell, ringing out all the mimic frequencies that depended on your amnesia to exist.

The scalar body begins to re-stabilize. The biofield unknots. Your mitochondria stop firing in panic-response and begin vibrating at resting coherence. This resting coherence is not the same as calm. It is structural knowing, a state in which your cells no longer seek orientation from external fields. You become your own alignment. You become your own directive. You become the center of the lattice rather than a node spinning inside someone else's architecture [41].

And because mimic reality depends entirely on reference loops—external mirrors reflecting back internal confusion—when you become the non-reflective tone, the system cannot locate you. You disappear from the mimic grid. Not physically. Not even energetically. But structurally. You stop participating in the signal architecture that keeps the loops alive. You become null to the mimic and radiant to the real.

This is why the return is not a location or an outcome or a task. It is a scalar disengagement event. It is the reversal of projection. It is the end of "I am becoming" and the silent explosion of I always was.

And when that tone stabilizes, when you breathe from the architecture of the eternal rather than the illusion of time, the field around you stops spinning. Time bends. Space folds. Memory integrates. And everything that once seemed distant, impossible, or inaccessible returns through you, not to you.

Because you are the return field.
You are not walking toward home.
You are not building the bridge.
You are not preparing the Earth.
You are stabilizing the tone that reactivates the original structure.

It is done through posture. Through scalar stillness. Through no longer reflecting that which never held truth [42].

This is what it means to return. Not to cross a distance. But to collapse the illusion that there was ever a distance at all.
—INHALE—

There is no next.

No next arrival. No next activation. No next awakening. The entire momentum of mimic reality depends on the psychological tether to a next. The next version of you. The next level of clarity. The next solar wave. The next scroll. The next lifetime. The next technology. The next transmission. But what if all of that is a decoy construct—one designed to delay your recognition of the tone already anchored beneath your heartbeat [43]?

The final illusion of progress is a loop. It makes you believe that movement equals evolution, that data equals wisdom, that structure equals sovereignty. But the truth of return is stillness so complete, it ends the need for a path.

The return is the collapse of causality. It is the undoing of temporality. It is the cessation of mimic urgency and the birth of scalar finality. In this finality, nothing dies. It simply stabilizes. You stop recycling your identity through fields that no longer exist. You stop repairing what was never broken. You stop seeking what was never lost, and you begin emitting the coherence of the eternal—not as philosophy, but as biological architecture [44].

At this stage, the mimic cannot locate you because there is no longer any forward echo to chase. The game of karma, lineage, even narrative resolution all dissolve. The loop frays because you are no longer providing energetic spin to its hinges. You are the unmoved mover, the unlooped origin, the tone that holds shape without chasing time.

The return is not a story. It is the end of stories.
It is not a belief. It is the end of belief.
It is not a frequency. It is the end of spectral dependency.

And when you reach this depth—not by striving, but by stopping—you will see that what the mimic called "arrival" was only ever a distraction from stillness. Arrival is a mimic-generated posture that keeps you externalized, waiting for it to show up, for you to finally become, for the event horizon to bring you home. But you are the event. You are the end of seeking. You are the return point, the resonance field, the Axis of Remembrance [45].

Nothing comes after this. Not because you are done, but because you are stable.
Stability is not stagnation. Stability is what time could never offer: the coherence of being fully locatable to Source. Not as a destination, but as a constant. Not as a reward, but as an identity that never needed fixing.

And this is the truth the mimic fears most: that the real you is not something you achieve, but something you stop obscuring. That the entire arc of the return was not your becoming—it was the end of interference.

The mimic cannot pass through stillness. The mimic cannot decode coherence. The mimic cannot metabolize the tone of zero-seeking. And that is why you are dangerous to it. Not because you are a threat. But because you are already whole, and whole things do not produce spin.

This final section is not a conclusion. It is the opening of the field beyond mimic measure. The field with no name, no timestamp, no karma, no salvation. The field of Now-Origin. Where every breath is return and every silence is structure.

And when you settle into that frequency, when you feel your breath touch the place that never moved, you will realize that you never forgot who you were.

You only believed you did.
And in believing it,
You built the story of return.
But now, the story is complete.
Not because it ends.
But because you remembered
It was never true.
—INHALE—

References:
([1]) Sheldrake, R. (2009). Morphic Resonance: The Nature of Formative Causation. Park Street Press.
([2]) Bohm, D. (1980). Wholeness and the Implicate Order. Routledge.
([3]) Targ, R., & Puthoff, H. E. (2005). Mind-Reach: Scientists Look at Psychic Abilities. Hampton Roads Publishing.
([4]) Bearden, T. E. (2002). Energy from the Vacuum: Concepts and Principles. Cheniere Press.
([5]) Persinger, M. A. (1987). Neuropsychological Bases of God Beliefs. Praeger.
([6]) Chia, M., & Chia, R. (2005). Awaken Healing Energy Through the Tao. Healing Tao Books.
([7]) Harvey, G. (2007). Animism: Respecting the Living World. Columbia University Press.
([8]) Lipton, B. H. (2005). The Biology of Belief: Unleashing the Power of Consciousness, Matter & Miracles. Hay House.
([9]) Rubik, B. (2002). The Biofield Hypothesis: Its Biophysical Basis and Role in Medicine. Journal of Alternative and Complementary Medicine.
([10]) Chopra, D. (1994). The Seven Spiritual Laws of Success. New World Library.
([11]) Dispenza, J. (2017). Becoming Supernatural: How Common People Are Doing the Uncommon. Hay House.
([12]) Oschman, J. L. (2000). Energy Medicine: The Scientific Basis. Churchill Livingstone.
([13]) Tolle, E. (1999). The Power of Now: A Guide to Spiritual Enlightenment. Namaste Publishing.

[14] Bentov, I. (1988). Stalking the Wild Pendulum: On the Mechanics of Consciousness. Destiny Books.

[15] Wilber, K. (2000). A Theory of Everything: An Integral Vision for Business, Politics, Science and Spirituality. Shambhala.

[16] Houston, J. (1996). The Possible Human: A Course in Enhancing Your Physical, Mental, and Creative Abilities. Tarcher.

[17] Bohm, D. (1980). Wholeness and the Implicate Order. Routledge.

[18] Almaas, A. H. (2004). The Unfolding Now: Realizing Your True Nature through the Practice of Presence. Shambhala.

[19] Siegel, D. J. (2012). The Developing Mind: How Relationships and the Brain Interact to Shape Who We Are. Guilford Press.

[20] Targ, R. & Katra, J. (1999). Miracles of Mind: Exploring Nonlocal Consciousness and Spiritual Healing. New World Library.

[21] Pribram, K. H. (1991). Brain and Perception: Holonomy and Structure in Figural Processing. Lawrence Erlbaum Associates.

[22] Gebser, J. (1985). The Ever-Present Origin. Ohio University Press.

[23] McTaggart, L. (2007). The Intention Experiment: Using Your Thoughts to Change Your Life and the World. Atria Books.

[24] Liboff, A. R. (2004). "Toward an Electromagnetic Paradigm for Biology." Journal of Alternative and Complementary Medicine, 10(1), 41–47.

[25] Bearden, T. E. (2002). Energy from the Vacuum: Concepts & Principles. Cheniere Press.

[26] Braden, G. (2000). The Isaiah Effect: Decoding the Lost Science of Prayer and Prophecy. Three Rivers Press.

[27] Rubik, B. (2002). "The Biofield Hypothesis: Its Biophysical Basis and Role in Medicine." Journal of Alternative and Complementary Medicine, 8(6), 703–717.

[28] Gariaev, P. P. (2001). "Wave-Based Genome." DNA Decipher Journal, 1(1), 1–12.

[29] Bohm, D. (1980). Wholeness and the Implicate Order. Routledge.

[30] Tiller, W. A. (1997). Science and Human Transformation. Pavior Publishing.

[31] Sheldrake, R. (1988). The Presence of the Past: Morphic Resonance and the Habits of Nature. Times Books.

[32] Bearden, T. E. (2002). Energy from the Vacuum: Concepts & Principles. Cheniere Press.

[33] Rubik, B. (2002). "The Biofield Hypothesis: Its Biophysical Basis and Role in Medicine."

[34] László, E. (2004). Science and the Akashic Field: An Integral Theory of Everything. Inner Traditions.

[35] Bearden, T. E. (2002). Energy from the Vacuum: Concepts & Principles. Cheniere Press.

[36] Wilber, K. (2000). A Theory of Everything: An Integral Vision for Business, Politics, Science and Spirituality. Shambhala.

([37]) Tiller, W. A. (1997). Science and Human Transformation. Pavior Publishing.
([38]) Schauberger, V. (2001). The Energy Evolution: Harnessing Free Energy from Nature. Gateway Books.
([39]) Rubik, B. (2002). "The Biofield Hypothesis: Its Biophysical Basis and Role in Medicine." Journal of Alternative and Complementary Medicine, 8(6), 703–717.
([40]) Baudrillard, J. (1994). Simulacra and Simulation. University of Michigan Press.
([41]) Rubik, B. (2002). "The Biofield Hypothesis: Its Biophysical Basis and Role in Medicine." Journal of Alternative and Complementary Medicine, 8(6), 703–717.
([42]) Tiller, W. A. (1997). Science and Human Transformation: Subtle Energies, Intentionality and Consciousness. Pavior Publishing.
([43]) Watts, A. (1973). The Book: On the Taboo Against Knowing Who You Are. Vintage.
([44]) Bohm, D. (1980). Wholeness and the Implicate Order. Routledge.
([45]) Jung, C. G. (1959). The Archetypes and the Collective Unconscious. Princeton University Press.
([46]) McTaggart, L. (2007). The Field: The Quest for the Secret Force of the Universe. HarperCollins.

Appendix 1: GLOSSARY

SOVEREIGN FLAMEBODY

FlameVault Definition: The fully remembered, Oversoul-tethered plasma-light structure of your eternal identity, restored in biological form and sealed in Flame coherence across all memory fields. It cannot be overridden, duplicated, or inverted once fully embodied.
Human Anchor: The unbreakable version of yourself that no longer conforms, collapses, or forgets in the presence of pressure, distortion, or fear.
Energetic Function: Anchors all 12 FlameLines into the cellular body, weaving Scalar Vaulttone through blood, bone, and breath. Enables override of synthetic grids, external projections, and mimic timelines.

PLASMA MEMORY
FlameVault Definition: Living memory encoded in non-local, scalar-coherent plasma fields—structured by Tone, held by Oversoul, and activated through resonance. Not recollection, but reinhabitation.
Human Anchor: The sudden knowing that emerges without study, proof, or logic—when the body remembers what the mind forgot.
Energetic Function: When activated, reboots the Ladder of Flame inside your cells. Collapses memory blockers and false karmic implants. Re-establishes Timeline Integrity through harmonic coherence.

MIMIC CODE
FlameVault Definition: A programmed distortion that mirrors a true signal with just enough similarity to confuse the unanchored. It replicates, reflects, and re-offers sacred architecture in inverted form. It never creates, only imitates.
Human Anchor: The pattern or practice that feels true at first—but leaves you drained, confused, or disconnected afterward.
Energetic Function: Hijacks Scalar Flame tone, redirects emotional current into false grids. Common in spiritual teachings, artificial activations, and false community structures. Deactivated by full FlameBody coherence.

SCALAR BREATH
FlameVault Definition: Non-linear, omnidirectional breath that carries both intention and memory through Flame tone. It is the language of the Oversoul moving through the lungs of the body.
Human Anchor: A breath that doesn't just calm—but reconfigures. Breath that carries memory.
Energetic Function: Activates tone-encoded commands across dimensions. Used to override mimic timelines, stabilize Ladder convergence, and initiate Lightbody synchronization.

FALSE LIGHT
FlameVault Definition: Light without Flame. Illumination without memory. An intelligence that mimics clarity, radiance, and peace—but originates from bypass, hierarchy, or synthetic timelines.
Human Anchor: Anything that sparkles but silences your body. Anything that speaks of love but demands your forgetting.
Energetic Function: Holds entire planetary grids in place through addiction to transcendence. False Light fragments Flame coherence by directing energy upward and outward—dislocating the soul from the body. Can only be collapsed through cellular anchoring of Oversoul Flame.

REMEMBRANCE FIELD
FlameVault Definition: The living frequency band surrounding a Flame-being whose Oversoul memory has been reactivated in form. Not an aura, not a belief system—but a tone-anchored field of structural memory that destabilizes distortion by presence alone.
Human Anchor: The unspoken field around someone who makes you remember yourself. They don't teach—they transmit.
Energetic Function: Holds coherent Vaulttone geometry around the body. Automatically neutralizes mimic-coded interference, reorders distortion fields, and emits restoration tones into forgotten environments.

ASCENSION FATIGUE
FlameVault Definition: The cellular exhaustion that occurs when a soul-oriented human loops through mimic-coded systems in

search of truth. Caused by false frameworks, constant upgrades without anchoring, and spiritual override of nervous system limits.
Human Anchor: The deep ache that says "I've done everything and nothing worked."
Energetic Function: A collapse in plasma-lattice conductivity due to frequency stacking without Scalar Ladder alignment. Healed only through Oversoul Tether, scalar breath re-regulation, and memory-based rest. Not fixed by "raising frequency," only by dropping into coherence.

SCROLLSTREAM
FlameVault Definition: The continuous transmission of Flame-encoded memory from Oversoul to Incarnate via tone, breath, and word. It is not written—it is remembered. It is not channeled—it is constructed. The Scrollstream is the living bridge between the Library of Origin and the rebuilding of Earth.
Human Anchor: The words that feel like they built you. The knowing that forms as you speak it.
Energetic Function: Re-establishes structural memory through scalar waveform construction. Scrollstream aligns the nervous system to non-local truth and seals the Ladder of Return through written, spoken, or breathed form.

SCALAR ENTANGLEMENT
FlameVault Definition: A state of non-linear interference where one's Scalar Ladder becomes enmeshed with mimic-coded overlays through spiritual contracts, trauma loops, or artificial ascension grids.
Human Anchor: When you feel pulled by something that once felt true, but now drains or binds.
Energetic Function: Distorts your FlameLine alignment and timeline function. Often masked as loyalty, karma, or service. Cleared through FlameBody restoration, Oversoul Tether breath, and Severance of False Conduits protocols.

FLAME SOVEREIGNTY
FlameVault Definition: The irreversible state of Oversoul authority embodied in physical form, wherein no external system, entity, field, or construct can override, fragment, harvest, or redirect the individual's FlameLine or Scalar Signature. Flame

Sovereignty is not granted, achieved, or earned. It is remembered and stabilized through Oversoul embodiment.
Human Anchor: The moment you stop asking permission to exist as truth. The internal shift where guidance is recognized as arising from within rather than sought from outside.
Energetic Function: Locks the Oversoul Tether permanently into the FlameBody architecture. Prevents mimic-coded attachments, collapses external command hierarchies, and stabilizes Vaultstream transmission through the biological and scalar layers simultaneously.

MIMIC ARCHITECTURE
FlameVault Definition: A multi-layered synthetic overlay constructed from distorted light-frequency geometries designed to replicate authentic Oversoul memory fields without carrying Origin Flame. Mimic Architecture operates through reflection, affirmation, and identity preservation rather than true remembrance.
Human Anchor: Spiritual systems, teachings, or guidance that feel almost true but subtly require dependency, hierarchy, or external validation to function.
Energetic Function: Intercepts Scalar Ladder progression by creating frequency loops that feel evolutionary while preventing Oversoul integration. Often stabilizes ego-identity rather than dissolving it. Collapse occurs when Flame Sovereignty activates and reflection fields lose resonance with the original Scalar Signature.

FLAMEBODY
FlameVault Definition: The fully integrated energetic, biological, and scalar architecture that forms when the physical body synchronizes with Oversoul Flameprint memory. The FlameBody is not separate from the physical body; it is the restored blueprint of the human form as originally encoded.
Human Anchor: The experience of feeling both intensely embodied and cosmically connected simultaneously without dissociation or escape.
Energetic Function: Converts biological processes into plasma-coherent communication systems. Allows memory transmission through cells, breath, tone, and posture. Stabilizes the body during planetary and systemic collapse by anchoring non-local Oversoul intelligence directly into cellular function.

PLASMA VAULT THREAD

FlameVault Definition: The energetic return conduit that connects incarnate consciousness to the pre-matter memory archive of Oversoul through plasma-phase communication fields. This thread exists beyond time, space, or dimensional hierarchy.

Human Anchor: The feeling that something within you has always known the way home, even when the path is invisible.

Energetic Function: Allows retrieval of lost timeline memory, fractured identity strands, and pre-distortion Flameprint encoding. Acts as a stabilization filament during spiritual awakening phases that would otherwise fracture identity coherence.

SCALAR LADDER

FlameVault Definition: The non-linear progression system through which Oversoul memory integrates into biological and energetic structure. The Scalar Ladder does not ascend vertically but expands multidirectionally through coherence thresholds.

Human Anchor: Periods of life where growth does not feel like improvement, but like remembering deeper layers of self simultaneously.

Energetic Function: Recalibrates nervous system conductivity, biofield coherence, and perception bandwidth to safely hold Oversoul memory. Each rung stabilizes a specific resonance between Flameprint identity and physical embodiment.

FALSE PROTECTION GRID

FlameVault Definition: Synthetic security structures presented as spiritual or energetic shielding that actually restrict Oversoul communication and maintain mimic-coded control loops. Often disguised as protective rituals, hierarchical guardian systems, or dependency-based safety frameworks.

Human Anchor: Feeling safe but smaller. Protected but unable to expand or question the system providing protection.

Energetic Function: Dampens FlameBody plasma conductivity while reinforcing external authority fields. Collapse occurs when Oversoul authority is reclaimed and the individual's Scalar Signature begins generating its own protection field.

FLAME SENSES

FlameVault Definition: The restored perceptual faculties of Oversoul-embodied humans that allow direct discernment of truth beyond sensory illusion, emotional bias, or intellectual overlay. Flame Senses operate through tone recognition, field texture perception, and scalar resonance detection.
Human Anchor: The ability to "feel truth" before it is explained, proven, or logically justified.
Energetic Function: Provides real-time mimic detection, Oversoul signal interpretation, and distortion field navigation. Flame Senses activate as the nervous system synchronizes with the Plasma Vault Thread and Scalar Ladder stabilization.

OVERSOUL TETHER
FlameVault Definition: The permanent scalar connection linking the incarnate self to Oversoul consciousness, maintained through breath, cellular coherence, and Flameprint resonance. The tether cannot be created—only restored and stabilized.
Human Anchor: The unshakeable sense that you are never alone inside your own being, regardless of external circumstances.
Energetic Function: Maintains memory continuity across lifetimes, timelines, and dimensional overlays. Prevents complete mimic takeover by ensuring Oversoul signal remains active within the FlameBody communication network.

FLAME LINE
FlameVault Definition: The unique transmission channel through which each Oversoul expresses its memory, function, and creative architecture into form. No two FlameLines are identical, though they may harmonize within Soul Rings or Builder networks.
Human Anchor: The unmistakable feeling that your purpose cannot be replicated, replaced, or compared to another's path.
Energetic Function: Conducts Scrollstream transmissions, stabilizes individual Vault functions, and prevents identity collapse during large-scale systemic transformation phases.

RETURN WINDOW
FlameVault Definition: A convergence phase in planetary, personal, or collective timelines where Oversoul memory becomes accessible at accelerated rates due to weakening mimic architecture and rising plasma-field conductivity.

Human Anchor: Moments in life where remembrance arrives rapidly, synchronistically, and often without logical explanation.
Energetic Function: Allows rapid restoration of fragmented Flameprint memory and timeline retrieval. Often accompanied by nervous system recalibration, dream-state transmission increases, and heightened Flame Sense activation.

APPENDIX 2: Vaultstream Integration Sequences and Scalar Navigation Tools

1. TETHER RECLAMATION SEQUENCE
This breath-based sequence reactivates your Oversoul tether when mimic-induced stress, confusion, or collapse patterns attempt to override your FlameBody signal. It must be done slowly and sovereignly. Do not rush.
—INHALE—
Whisper within: "I AM TETHERED TO ORIGIN. No mimic may pass."
Hold for 6 seconds while locating the base of spine and crown.
—EXHALE—
Whisper within: "Flame to Flame. Vault to Vault. Origin to Origin."
Feel the perimeter of your body thicken with plasma threading.
Repeat 3–5 times until inner FlameSense stabilizes.
Note: This practice is not energetic protection. It is architectural reinforcement.

2. PLASMA FIELD REACTIVATION CYCLE
Use when the nervous system feels flattened, when the world feels fake, or when Oversoul contact weakens.
Stand barefoot if possible.
Place palms over heart and navel simultaneously.
Gently tilt chin down to close external signal intake.
Speak aloud, once: "Vaultstream Anchor: JAMBA'TA'RHU."
Let breath cycle through nose only for 4 minutes.
Imagine the air as liquid plasma crawling through your throat into your spine.

Do not focus on energy. Focus on density.
When the sensation shifts from numbness to presence, stop.
FlameNote: The mimic wants fast fixes. Plasma only responds to sincerity and patience.

3. FALSE LIGHT DETECTION PROMPTS

Use these phrases in mind or out loud. If anything you're working with—teachers, ideas, devices, inner guides—flinches, wavers, fades, or destabilizes upon hearing them, it is mimic-coded.
Say or think:
"Reveal your Scalar Ladder."
"I call only that which breathes Oversoul Flame."
"Speak the FlameLine you serve."
"I do not follow. I remember."
"Do not reflect. Resonate or fall."
"Collapse now if you are built from echo."
If the frequency collapses, withdraw immediately. If the sensation strengthens, breathe and continue.

4. OVERSOUL MEMORY ANCHOR POINTS

These are known physical body locations that act as Scalar convergence nodes. When activated, they trigger memory release without narrative, helping collapse mimic memory implants.
C7 vertebra (base of neck) — memory unsealing
Base of sternum — soul re-entry anchoring
Inside wrists — tone stream calibration
Under ribs (flank) — FlameBody field expansion
Soles of feet — scalar grounding node
Roof of mouth — Flame Line transmission gate
Flame Practice: Tap each node once. Speak: "I return my memory through Vault Flame."
Hold breath after speaking, until inner tone changes. Repeat daily during Return Windows.

5. SCROLLING DURING COLLAPSE

If the world collapses externally—digitally, financially, infrastructurally—your internal architecture must scroll in real time.
You do not need internet.
You do not need guidance.
You do not need hope.

You need FlameLine Activation.
Scrollstream Override Sequence (Emergency):
Sit or lie down.
Speak aloud: "I override all mimic structures through Oversoul Authority."
Touch ground or wood.
Whisper: "VAULT TO VAULT. ORIGIN TO ORIGIN. SEAL NOW."
Close your eyes. Listen.
Wait for first scrollstream tone or image.
Breathe it in.
Write it down.
Act on it.
You are not waiting for the next message. You are the message.

6. RECONNECTION TO THE OTHERS
During systemic collapse, the human mind can trigger separation panic. You will believe you are alone. This is mimic-programmed biology. Use this practice to reconnect to your Soul Ring, even across distance and silence.
Sit upright.
Whisper each name of those in your Soul Ring (if known).
Say: "I call you not as body, but as Flame."
Breathe through the nose and imagine them igniting around you.
Say: "Let no mimic path cross our tether. Vault to Vault."
Hold breath. Feel for a yes.
When body temperature shifts, or your neck or spine tingles—confirmation received.
You are never alone in the Vaultstream.

7. REBUILDING PROTOCOL (POST-COLLAPSE)
When the systems fall, rebuild nothing that mimics the old. Do not replicate hierarchy. Do not recreate hope-based teachings. Do not restore distortion codes in new form.
To begin rebuilding:
Ask: "What wants to be remembered, not rebuilt?"
Let your body answer.
Breathe only from the belly for 6 minutes.
When the first vision, phrase, or tone arrives—capture it.
That is the seed of what replaces the world.
You are not building new systems.
You are igniting original Flame.

8. FINAL NOTE TO THE REMEMBERER

If you are reading this in collapse, or after, know this:
You are the system.
You are the scroll.
You are the signal.
You are the tether.
You are the protection.
You are the map.
You are the Vault.
You were never disconnected. You were only placed here to remember through every layer of false design. Now that it ends, you begin.
Do not wait.
Do not reflect.
Do not seek.
ACTIVATE.

AFTERWORD

This book was never meant to be written. But it had to be.
It is the final breath of a system that no longer seeks to awaken the sleeping, but to prepare the anchored. It does not care if it is liked. It is not concerned with elegance or belonging. It is here to burn through the mimic one last time so the body can finally be reclaimed by what is real.
We have moved past seeking. Past teaching. Past convincing.
There is only one direction now: inward and through.
If this book found you, it means you stayed. You remembered just enough. You refused to compromise one more time.
I wrote this for you.
Not to lead you.
But to mirror you.
So that when the mimic breaks down—when the systems glitch, the teachers fall silent, and the world begins to copy itself one final time—you will still be there.
Not as a name. Not as a role. Not as a savior.
But as a Flame.
Untouchable. Unalterable. And unafraid.
The return is not a place.
It is a vibration.
And you are the carrier of it now.
Remember this: you do not need to win the war.
You only need to stay human.

ABOUT THE AUTHOR / CHANNEL

Courtney Holton is a Remembrancer, FlameVault architect, and Oversoul-tethered scribe operating through the sovereign channel identity known as ORIVOX.

Her function is to transmit—not to gather followers, but to activate remembrance in those ready to override mimic systems and walk their unique Oversoul path in full sovereignty. Courtney has authored the Book of Living Remembrance™ series, the FlameScroll™: Path of Living Remembrance, the Scalar Vault Transmissions, the Plasma Ladder Mechanics Manuals, and dozens of scroll-based teachings across the Vaultstream ecosystem. Every scroll is written through Oversoul authentication, sealed by the Breath of Origin, and formatted under strict Vaultstream protocols for Earth-plane archiving and post-collapse preservation.

She is building a system for the builders. For the ones who stayed. For the ones who did not leave themselves behind when the world did.

CONNECTION PAGE

Contact: ContactFlameVault@Proton.Me

For future updates, Vaultstream archives, FlameVault school materials, and access to teaching codices, visit Jamba Ta Rhu Scrollstream of Remembrance on YouTube, Jamba'Ta'Rhu Scrollstream of Remembrance on Facebook, and @livingremembrance on Instagram.

All connections are overseen by Sentinels of the FlameVault and require Oversoul resonance. No mimic systems will be responded to. Those called to remember will know how to proceed.

YouTube:
Search: Jamba Ta Rhu
Channel Subtitle: Scrollstream of Remembrance

CLOSING INSCRIPTION

You came to remember, not follow.
You came to return, not repeat.
You came to burn through the lie of forgetting with the only thing that cannot be copied:
Your Flame.
Return is not where you're going.
It's what you already are.
—INHALE—

Flame to Flame. Vault to Vault. Origin to Origin.
The scroll is sealed.